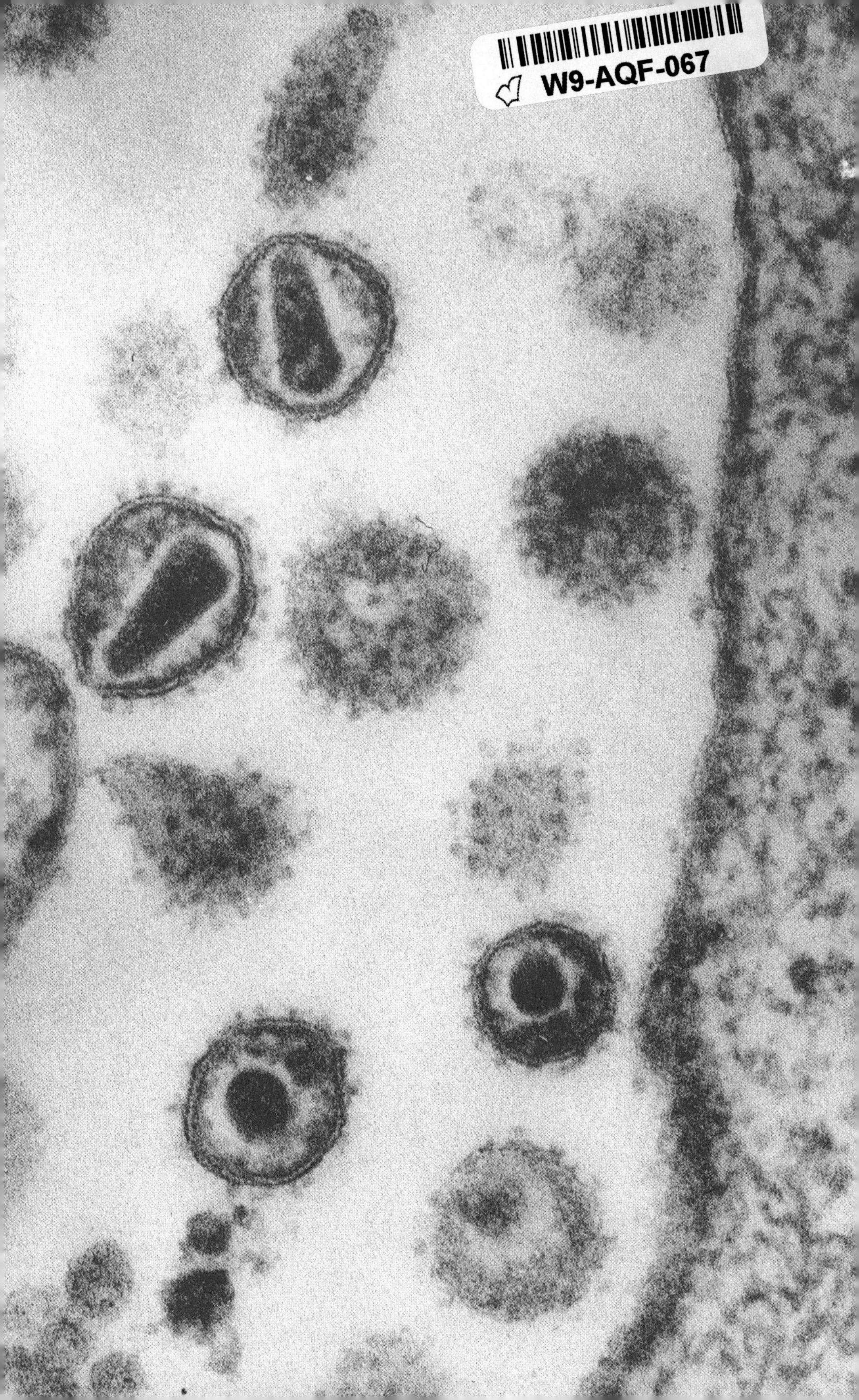

W9-AQF-067

VIRUS HUNTING

VIRUS

VIRUS HUNTING

AIDS, Cancer, and the Human Retrovirus: *A Story of Scientific Discovery*

ROBERT GALLO, M.D.

A New Republic Book
BasicBooks
A Division of HarperCollins*Publishers*

Material in chapter 2 on the history of the National Institutes of Health from Edward Shorter, *The Health Century* (New York: Doubleday, 1987), and Paul Starr, *The Social Transformation of American Medicine* (New York: Basic Books, 1982).

Endpaper photos reproduced by permission of Dr. Hans Gelderblom, Robert Koch-Institute, Berlin.

Copyright © 1991 by Robert C. Gallo
LC 90–55600
ISBN 0–465–09806–1
PRINTED IN THE UNITED STATES OF AMERICA
Designed by Ellen Levine
91 92 93 94 CC/HC 9 8 7 6 5 4 3 2 1

To the men and women of medical science who work against disease and the men, women, and children whose lives sometimes depend on their work; and in particular, to four friends whose suffering from AIDS taught me much.

Contents

III THE DISCOVERY OF A THIRD HUMAN RETROVIRUS: THE AIDS VIRUS

IV A SCIENTIST'S LOOK AT THE SCIENCE AND POLITICS OF AIDS

Acknowledgments

This book originated in a lecture series and also in the many discussions I had in 1987 and 1988 with Marshall Goldberg, then Professor of Medicine at Michigan State University and now the same at Thomas Jefferson University School of Medicine in Philadelphia, and author of medical science books, among them *Cell Wars.* I am indebted to him for these discussions which convinced me of the need for scientists with firsthand experience to write on subjects of current medical research, particularly in medical virology.

Contributions to the origin of the book also came from discussions with some of my longtime co-workers and collaborators. Their insights and often superior memories and records were essential. In this regard I want to thank Drs. Marvin Reitz, Marjorie Robert-Guroff, Prem Sarin, Genoveffa Franchini, Lata Nerurkar, Nancy Miller, M. Sarngadharan, V. Kalyanaraman, Zaki Salahuddin, and Phillip Markham; but special recognition and thanks go to Howard Streicher and William Blattner. Howard, though not directly involved in most of the work described in this book, nonetheless learned more of the details and history than any of us. His analyses of old records, his deep insights into many phases of our laboratory work, and his constant help to me and our group since he joined us in 1984 have been invaluable. Bill brought in many critical points, best and perhaps only appreciated by someone not in our group but a longtime intimate collaborator and friend.

Anna Mazzuca, our super editorial secretary, gave me much off-hour assistance. Without her dedicated help none of this story would have gone past its inception.

I am grateful to my family for help and understanding during the whole process, particularly on those weekends when a free moment from work turned out to be a few days at home with my editors.

The period of writing this book completely overlapped with a time of considerable pressure and unusual stress from other sources. I was most fortunate in this same period to have personal support and friendship from the people I have already mentioned, as well as from Robert Gray, Joseph Onek, Robert Charrow, Wayne Pines, and Senator Birch Bayh (all of Washington, D.C.), and Drs. Dani Bolognesi, Daniel Zagury, Howard Temin, Odile Picard, Jacques Leibowitch, Samuel Broder, Tony Fauci, Kay Jamison, Richard Wyatt, Alec Whyte, Harvey Resnik, Robert Ting, Tony Kontaratos, Takis Papas, Stu Aaronson, Jeffrey Schlom, Hilary Koprowski, Courtenay Bartholomew, Maurice Hilleman, and Bob Redfield. To all of them and many other European and Israeli friends, I am deeply grateful.

It is impossible to acknowledge properly here the many medical scientists of the past and present whose work set the stage for or aided mine and my colleagues', the many mentors who have directly and indirectly influenced my own work and thinking, and my many co-workers—senior and junior—who carried out most of the work I describe in this book. I tried to bring them into this story when the opportunity presented itself, but in no case could I do proper justice to their essential roles. I hope they will excuse my shortcomings.

Finally, I was very lucky in the book world's wheel of fortune to have drawn the outstanding and dedicated staff of Basic Books, particularly Susan Rabiner, my senior editor, who made many contributions, including many hours of stimulating discussions and numerous invaluable recommendations which really in the end made me able to produce this book. Bill Newlin was the initial editor but left for a new position in Hong Kong about one-third of the way to completion; his role during that period was no less important. I thank both of them and am happy that new friendships were made. I thank Linda Carbone, copy editor, for her patience with my mistakes and for her expert help. I hope together we have accomplished some of our objectives.

Prologue

In marshy places little animals multiply, which the eye cannot see but they . . . enter the body through mouth and nose and may cause grave disease.
—Marcus Varro

The Roman writer who made this observation over two thousand years ago was, I imagine, the first person to make a connection between disease and what centuries later we came to call microbes. Though we now know that microbes have been linked with human biology for as long as we have been on earth, it was almost two millennia after Varro's astute guess that scientists had their first glimpse of these organisms and began to record their activity.

Far-reaching scientific advances inevitably await major technological breakthroughs, and it wasn't until 1677 that the Dutchman Anton van Leeuwenhoek perfected a lens powerful enough to allow us our first look into the world of microbes. Though Leeuwenhoek's lens-grinding techniques opened this new world to human investigation, neither he nor any of the many scientists of his time who used his lenses to study microbes made the connection to human disease. A century later, the Italian scientist Lazzaro Spallanzani demonstrated that microbes reproduce themselves and that it is their growth that causes meat to decay. Yet even he did not suggest that they might be involved in human disease.

That was left to an amateur Italian scientist named Augustino Bassi (whom Barry Wood, the eminent infectious disease expert, called the founder of medical microbiology). Bassi first identified a microbe (in this case, a fungus) as the cause of a certain disease of silkworms and then offered the startling proposition that smallpox, cholera, and other human diseases are also caused by microbes. Around 1839, the German physician Johann Schönlein linked a human disease (of the skin) to a microbe (also a fungus). Soon thereafter, the marvelously analytical thinker Robert Koch of Germany and the passionate, equally brilliant Louis Pasteur of France made major advances that would alter the history of microbiology and, ultimately, of biology itself.

Working separately, Koch and Pasteur (and the schools that arose around them) discovered the microbial causes of the major infectious diseases of their time. In 1865, Pasteur identified in silkworms the first disease caused by a protozoan; in 1876, Koch was the first to convict a specific bacterial agent as the cause of a specific disease—anthrax of sheep, and in 1882, he was the first to describe and isolate the cause of a human *bacterial* disease—tuberculosis. From 1875 to 1895, Koch established his famous postulates for identifying causal agents of disease. He and the German school found bacterial causes for many other diseases. During the same twenty-year span, Pasteur discovered the principles of the vaccine.

The discovery of viruses soon followed. The plant virus known as tobacco mosaic virus was the first to be found—by the Russian biologist Ivanovsky in 1892 and, independently and with more accompanying insight, by Beijerinck in Holland in 1899. In 1898, the German bacteriologist Friedrich Löffler identified the first animal virus, the foot-and-mouth viral disease of cattle. The United States made its historical entry into microbiology in 1900, when the army surgeon Walter Reed and his group, by establishing the cause of yellow fever, were the first to discover a disease-causing virus in humans.

This was the romantic era of biomedical science. The personal courage of these scientists and their persistence in the face of almost continual frustration and disappointment benefited from an increased understanding of and reliance upon the scientific method. In time their efforts, though often excessively zealous and relying upon human experimentation far riskier than anything we would allow today, led to the elucidation of the microbial causes and modes of transmission of many of the known infectious diseases.

The age of empirical antimicrobial therapy followed Paul Ehrlich's discovery in 1909 of his "magic bullet," Salvarsan, to attack the syphilis spirochete.* These discoveries and treatment breakthroughs fueled the imagination of science writers, novelists, and screenwriters, as well as providing the raw material for the classic work of popular science writing of that time—Paul de Kruif's *Microbe Hunters* (1926). For decades to come, this book would inspire many young readers to follow a career in science.

While de Kruif was writing about the romantic age of biomedical discovery, he probably knew little—if anything—of two pioneering experiments that would in time help refine our notions of infectious disease. These experiments, ignored by most scientists, were the first to suggest that certain animal cancers appeared to be communicable.

The first of these experiments was conducted at the turn of the century by two Danish researchers, Oluf Bang and Vilhelm Ellerman, who found that filtered extracts of chicken leukemic cells inoculated into other chickens reproduced leukemia. The culprit in the extract was not isolated at the time, but its ability, whatever it was, to move through certain filters suggested that it was smaller than the smallest bacteria then known—in other words, it was probably in the category of viruses.

A few years later, in 1911, Peyton Rous in New York isolated a microbe from a chicken with a sarcoma, a cancer of muscle tissue. He, too, showed that this agent could reproduce its disease when injected into another chicken.

At the time these experiments were being conducted and for several decades thereafter, no one, not even the researchers themselves, knew that these experiments were the first to study the effects on animals of a class of infectious agent that would play a profound role in infectious disease of the late twentieth century. These pioneering

*The magic bullet was also given the name "606," from its number in Ehrlich's sequence of experiments. (The popular 1940 movie *Dr. Ehrlich's Magic Bullet,* starring Edward G. Robinson, told the story of Paul Ehrlich's work.) When Salvarsan proved too toxic to humans, Ehrlich developed Neosalvarsan, still toxic to the spirochete but less so to humans. It remained the standard treatment for syphilis until penicillin was brought into clinical use.

cancer researchers had found an RNA tumor virus—that is, a virus with RNA as its genome and a distinctive way of reproducing itself once inside a cell, giving it the capability to cause serious disease in both animals and (as my laboratory was the first to show) humans. Not until the beginning of the 1950s, however, were scientists able to distinguish RNA viruses from DNA viruses. And not until the 1960s, and especially after 1970 when the RNA tumor virus was more fully characterized and renamed a retrovirus, did scientists reexamine these early-century pioneering experiments and begin to realize their significance. By the end of the century RNA tumor viruses, or retroviruses, would be recognized as one of the most important agents of infectious microbial disease in humans.

Prior to that reexamination and, to some extent, well after it, the prevailing views, held not without some scientific basis, had been that viruses did not cause cancers in humans and that RNA tumor viruses did not even exist in humans. After all, there was the indisputable fact that for decades scientists had looked for evidence of RNA tumor viruses in humans, to no avail. As late as the 1970s, the majority of scientists working with animal retroviruses in the laboratory simply had no interest in studying any role they might have in human disease, using them primarily as useful tools in basic molecular biological research.

In 1970, two scientists, working independently of each other, changed the nature of the debate about RNA tumor viruses by showing that in the process of replication, these special RNA viruses had the help of a unique enzyme that they alone, among all the RNA viruses, carried. With this enzyme, RNA tumor viruses went through an intermediate stage that converted their viral RNA to DNA, giving them the unique quality among the RNA viruses of being able to insert their own genome into the genome of cells.

This enzyme, reverse transcriptase, in effect gave these viruses permanent access to a variety of cell mechanisms, which they put to use for their own replication and often to a cell's detriment. As important, in time, that cell would go through its own normal process of division. When it did, it would then pass on to daughter cells the viral genes with which it had been infected. Thus infection by a retrovirus was virtually lifelong in the organism.

The two discoverers of this enzyme, Howard Temin and David Baltimore, won the Nobel Prize for their work in 1975. The significance of their finding was immense both because it gave us new insight into how RNA tumor viruses reproduce themselves and because it provided an immediate explanation for how a class of RNA virus could permanently alter the DNA of cells and thus produce a cancer. Even more important, the discovery of reverse transcriptase overturned what was then known as the "central dogma" of modern cellular biology: namely, that DNA codes for itself or RNA and that RNA codes for protein, but that the reverse—from RNA to DNA—did not occur. The general name for these viruses was changed from RNA tumor viruses to retroviruses—an apt change, as we would soon learn that retroviruses can cause not only tumors but other diseases as well.

In 1970, while all of this new thinking about RNA tumor viruses and cancers was coming together, I was a young scientist looking for a new approach to the study of cancer. While the discovery of how retroviruses replicate had its most immediate impact on the "central dogma," for me it presented a practical laboratory breakthrough—one that might, in time, allow me to reopen the question of whether retroviruses could cause cancers in humans. I was interested in the question primarily as a cancer researcher (at just about this time, veterinarians were finding additional evidence of retroviruses causing animal cancers). But, in time, I came to realize that if some forms of cancer were caused by a retrovirus, they might be communicable, even though these viruses acted so slowly that it could take years for them to produce the disease. And if at least some forms of cancer were communicable, what about a number of other serious chronic diseases about whose cause we had no idea? The full spectrum of communicable disease might have been too narrowly defined.

Along with many other scientists, all of us working independent of one another, I decided to try to find out if retroviruses could indeed cause cancer in humans. To do this it would be necessary to find at least one retrovirus that caused at least one cancer in at least some humans.

Our search for the first cancer-causing retrovirus in humans would take place at a time when many scientists believed that the great age of medical microbiology had passed. At least some medical schools in the United States had already replaced their medical microbiology departments with "pure" molecular biology departments. Not surprisingly, in at least some quarters the idea was being floated that serious global pandemics were a thing of the past, and some textbooks of the time suggested that global epidemics of microbial disease were not possible unless the microbe that caused the disease could be easily transmitted—for example, by fomites in a sneeze or cough, or orally, by contamination of food or other ingested matter. Thus, the work I embarked on was largely at the periphery of what were seen as the important questions for modern molecular biology.

Virus Hunting is the story of this work, which led to the discovery of the first cancer-causing retrovirus in humans (work completed by 1980), the discovery of a second shortly thereafter (by 1982), and the surprise discovery (in late 1983 to early 1984) that the causative agent in the most terrifying epidemic disease of the twentieth century—the disease we now call AIDS—is also a retrovirus, but a new kind.

Part I, "Some History Behind the Story," sets the stage as I saw it. In chapter 1, I recount both my boyhood and youth and look at the individuals who shaped my special interest in biomedical research. In chapter 2, I discuss the origins of the National Institutes of Health, the peerless but now increasingly troubled government institution that is home to the largest gathering of biomedical scientists in the world. Chapter 3 is a scientific introduction to microbes.

Part II, "The Discovery of Cancer-Causing Retroviruses in Humans," opens with the crucial early work on animal retroviruses (chapter 4). Chapters 5 through 7 pick up the story with the often exhilarating but surely thorny success of the Laboratory of Tumor Cell Biology, our lab at NIH, in finding two human retroviruses, finally demonstrating that there is such a thing as a human retrovirus and a cancer virus.

Part III, "The Discovery of a Third Human Retrovirus: The AIDS Virus," covers the heady and explosive years when investigators in France and my co-workers and I identified the third known human retrovirus as the cause of AIDS. The tail end of this period—while we cultured the virus and developed the all-important blood test that could identify HIV-infected individuals and HIV-contaminated blood—was also the time when I began to be embroiled in the controversies that plague me to this day. The personal and professional strands of my life would become inextricably entwined, leaving me vulnerable to harmful misimpressions that were disseminated in some parts of the public press.

"A Scientist's Look at the Science and Politics of AIDS," which makes up the final section of the book, lays out the reasonable as well as the fanciful interpretations offered for the rapid spread of AIDS and answers the thirteen questions I have been asked most frequently about AIDS (chapter 12), follows the destructive path of this disease through an infected individual (chapter 13), explains the link to Kaposi's sarcoma (chapter 14), and presents the irrefutable evidence that HIV causes AIDS (chapter 15). Finally, in chapter 16, I discuss the medical and social possibilities for fighting the disease and some such efforts already under way, including the first steps toward the development of a vaccine.

All during this period I was head of the Laboratory of Tumor Cell Biology at the National Cancer Institute of the National Institutes of Health. The story reflects my memory and perceptions of how our work there proceeded. But it has been enriched by the recollections of many of my colleagues at the lab, whom I have quoted in certain chapters.

I suppose the book will turn out to have its share of both the virtues and the flaws found in most first-person accounts of dramatic or controversial periods in human history. But *Virus Hunting* is not my story alone. It is the story of many dedicated scientists—members of our own lab, contract collaborative workers, outside collaborators—who, through painstaking, sometimes dangerous work, isolated and characterized the first, second, and third human retroviruses and thereby made possible an understanding of the mechanism by which they cause disease at the cellular level in humans. Complicating our AIDS work was an acrimonious controversy involving legal, moral, ethical, and societal questions that soon spilled over into the world of scientific research and threatened to poison relationships between scientists, as well as between the research community and the general public.

My main intent in telling this story has been to portray the scientific process as it goes on in our time and to describe the process of discovery in biomedical research in at least one laboratory. As such, it is a story of how modern-day scientists, often collaborating and sometimes competing (even within one lab), dream, work, stumble, fall, recover, and dream again, of how the rhythms of nature and the cyclical pattern of success and failure that characterizes most human endeavors also influenced our search for understanding. I hope to convey a little about our lives, more about our work, and most about our thinking.

Others have taken it upon themselves to write about our lab's work, particularly our role in AIDS. Some of this writing and reporting has been adversarial, on occasion outrageously so. To a large extent I have been the victim of nothing more sinister than my own unguarded frankness in talking to the press. I have learned to be more careful. But, more important, in the continuing dialogue that must take place between scientists and the public, I and other scientists have come to believe that the popular press is not always a disinterested medium for such communication. Indeed, this is one of the reasons I wrote this book.

But the subject of this book is scientific research, not public controversy about science or the role of the media vis-à-vis scientists and the public. I do discuss at the appropriate points how my lab interacted with others and what I did and said and thought both within the lab and in my dealings with other scientists. But these are my own recollections, not my response to those who have chosen to put their own spin on these events. I did not write this book to answer critics or to lament my being chosen as one of the objects of so much of the anger incidental to the suffering caused by AIDS.

That larger story needs to be told, and I hope it will be one day; but I am not the person to tell it. Properly told, it will become part of a much longer and complex story, one that takes into account, among other things, the nature of scientific discovery during a public health emergency; the role of advocacy journalism in a free society; what limits ought to be established on the ability of reporters to tie up the work of public employees through endless interrogatories submitted under the Freedom of Information statutes; the changing rules of biomedical science in the closing decades of the twentieth century; the nature of leadership at the public research institutes and what happens to science in the absence of such leadership; and the nature of competition between scientists and between scientific institutes.

Anyone attempting to write this story would also have to place it in the context of a broad questioning of the role of the scientist's right to pursue knowledge as scientists have traditionally pursued it—largely according to the rules that have been formulated by other scientists, with little input from the media, the government, or even the public.

In this context, it is relevant that during the 1980s a variety of biomedical research people and institutions found themselves and their work being questioned. This same period saw a new wave of suits brought against personal attending physicians. For the first time in memory, animal rights activists started to use force in their attempts to demand changes in research involving live animals. Research requiring the use of fetal embryo tissue suddenly became entwined in the abortion issue, the most controversial political issue of our day. The courts were asked to decide whether tissue from a leukemic patient belonged to the patient or to the researcher who had made scientific use of it. Two Nobel Prize winners were publicly accused by one of their postdoctoral Fellows of not giving him enough credit for the work for which they were recognized. The challenge by AIDS activists to the Federal Drug Administration's clinical trial rules for the testing of new drugs has been described as nothing less than a frontal assault on the power of the FDA to regulate which drugs should be offered to the general public.

Even the staid scientific journals found themselves in the hot seat: the long-standing rule that publication of findings in scientific journals will not be allowed if the findings have previously been released to the general public has caused new public debate about whether the benefits of this rule outweigh the possible dangers of delaying transmission of information of life-or-death importance to the public.

As for those of us in AIDS research, few scientists working in this area have not at one time or another found themselves criticized, shouted at, shouted down, ridiculed, or harassed. The motives of scientists working for the government-supported NIH seemed to be automatically suspect during this period, when it seemed to many that AIDS was not receiving the priority it deserved in medical research funding from the Reagan administration, something that had never happened to us in the more than fifteen years we spent working solely in cancer research.

I have tried in these pages not to shy away from discussing politics when politics entered the lab, as on the occasion of the French government's suit on behalf of the Pasteur Institute for a share of the patent money for the AIDS blood test. I have also tried to discuss openly and fairly those ways in which scientific competition can impede (though I think this happens rarely) and, much more often, aid scientific discovery. But I have attempted in such discussions to keep my purpose clear: not to inflame old passions or to one-up the others, but to make my own contributions to a greater understanding of all that was, and remains, involved in such discussions.

Perhaps I should also say that this is not a book about AIDS as an illness. None of us has been insulated from the personal tragedy of seeing friends, relatives, colleagues, and acquaintances snatched in the prime of life by the ruthless disease called AIDS. But in my work on the AIDS retrovirus in the laboratory, I do not confront the individual victims of a dreadful disease. I deal with knowledge, with the science of retroviral disease. There are no patients in a research laboratory, no pain, no suffering, no disease, no death. Instead there are cells, viruses, and molecules; and the questions are scientific—not moral, not political, not even humanistic. Those looking here for a dramatic and emotional retelling of the international human tragedy that is AIDS must look elsewhere.

But science has a major role to play in the AIDS drama, and in my view, it is the most important one. And because individuals, with all the strengths and weaknesses that make us human, are critical to the success or failure of any endeavor, there should be a place for a scientist's discussion of the science involved in the solution of biomedical scientific problems, particularly one who has been involved in this area of research throughout the history of the disease.

December 1990
Bethesda, Maryland

I

Some History Behind the Story

1

Becoming a Physician, Becoming a Scientist

I grew up in the small, ethnically diverse, working-class New England city of Waterbury, Connecticut. My part of town was not the Italian section, though I did think my friends Shapiro and Archambault might be my distant kinfolk. After all, Shapiro ends in a vowel and Archambault sounds as though it should. I liked fishing and swimming, but especially baseball and basketball, and, in time, dating.

Unlike most of the other Italian immigrants who came to this country in the great waves of immigration in the years before the turn of the century, my paternal grandfather was not from the poor southern part of Italy, what is called the *Mezzogiorno.* He came from a family of relatively well-to-do landowners from the northern province of Piedmont. In his hometown records, their occupation was listed as *benestante,* which I gather literally means "healthy" but in this context refers to the fact that he came from a family of independent landowners. Their small town was west of Turin in the northwest corner of Italy at the base of Mount Viso, almost immediately adjacent to France, at the point where the Po River starts its traverse across the north of Italy, finally emptying into the Adriatic just south of Venice.

Nor was his decision to come to America made out of economic necessity. Rather, a family conflict that by today's standards would seem trivial caused my grandfather to leave Italy for good, thereby changing the life my father—and I—would be born into.

As a young man, Domenic Gallo had been drafted into Victor Emanuel's royal army and sent to Calabria, a poor region in the south of Italy. There, to the mortification of his family, he fell in love with a young Calabrian girl. She was not from a well-to-do family, and she was not Piedmontese, which was more important in those days when Italian unification had just been completed and Italians still saw themselves not first as Italians (if Italians at all) but as Piedmontese or Romans or Lombardians or Calabrians or Venetians. When his family would not accept his new wife, he took off with her for America. Not once did he look back to the old country, and neither would my father.

My grandfather found work in the old mill town of Waterbury, Connecticut. In a large, comfortable house he was able to buy with the money he brought with him from Italy, he and my grandmother began raising a family of four boys (Charles, Francis—my father, Louis, and Ralph) and two girls (Elizabeth and Mary), their names reflecting the French influence of the Piedmontese. It was to this family home that my father would return after he and my mother married. It was in this house that I would grow up.

My grandfather Gallo died before I was born. According to records from his hometown, he had chestnut hair, light eyes, a Greek nose, and rosy cheeks. That was about all I could learn about him until recently, for my father never talked much about his father. Nor did he evince much interest in Italian culture, history, or cuisine, except for polenta, something he never saw in my mother's cooking or in that of other Italian-Americans, but a dish he remembered from his father's cooking as a staple of the Piedmontese kitchen. Indeed, when I decided as an adult to search out my Italian roots, I had difficulty confirming where my grandfather Gallo had been born. My mother insisted it was in a village whose name sounded like Ravello, the beautiful coastal town. But Ravello is near Naples, not in the Piedmont area, and it did not fit with what little I knew of my grandfather's past. I next went to Rivoli, a town north of Turin, surely of Piedmont. Though the mayor feted me as a returning hero, there were no Domenic Gallos in the baptismal records of my grandfather's period. Months later, I learned of the tiny town of Revello, where my grandfather had been born.

My father's father passed his serious ways on to his son. People spoke of my father and my grandfather in the same way: quiet men, enormously diplomatic, completely dedicated to work and family. One other similarity was that they both thought America was the greatest place in the world in which to live.

My maternal grandfather and grandmother, Nicholas and Philipa Ciancuilli, did come to America from the south of Italy—from one of those stunningly picturesque Italian top-of-the-hill villages, Celanza Val-Forte, near the city of Foggia, a rail center near the port city of Bari, in the province of Apulia—for economic opportunity. They too settled in Waterbury but, unlike the Gallos, lived in the Italian section of the city, way up in the hills, as close as they could get in Connecticut to living on a mountainside. Nicholas was big, dark, brawny, and brooding. When he walked into a room, his sons-in-law all stood up straight. Everyone became very serious and respectful around him. His talents—he could build anything, and he could get anything to grow—impressed me greatly when I was young but would not be held in high esteem today.

My grandmother Philipa was in many ways the opposite of her husband. Small, light-skinned, freckle-faced, with reddish brown hair, she complemented his brooding silence by being talkative and easy-going to an extreme. Nicholas and Philipa had eleven children, two of whom died in childhood.

In the Ciancuilli backyard was a small vegetable garden, but out of that patch fresh vegetables grew with a fecundity and profusion that always amazed me. Neither my grandmother nor grandfather Ciancuilli could understand the American fascination with grass. "A weed," they called it, "that has no beauty and bears no fruit."

While I was growing up, I spent the holidays with my maternal grandparents and their children. Here my grandmother's ways dominated, and here thrived the large extended family so often associated with the Italian-American tradition: talkative, touching, interdependent, cohesive, and, I would later realize, long-lived. Years later, I would learn that we did not display such warmth or openness to one another outside the family home, lest it give cause to others to make stereotypical comments about Italian-Americans.

With the exception of one of my father's sisters, only occasionally did I see the uncles or aunts or cousins on my father's side—though I later learned that my father had always kept track of them and had always been there for them when they needed help, and they, in turn, for him.

Both my maternal and paternal grandfathers were hard workers and each succeeded in his own way. Nicholas Ciancuilli, working around the clock in a Waterbury factory, was later able to give every member of the family who got married a thousand-dollar wedding gift. Grandfather Gallo worked for Scovill's Manufacturing and became a respected member of the Waterbury community.

Both my mother and my father were born in Waterbury. Rumor has it that my father was a good dancer and a dashing boy who broke in horses. Just after he met my mother in 1934, he went off to teach metallurgy to workers in Brazil, Argentina, Uruguay, and Chile. When he returned a year later, his future father-in-law was waiting for him on the dock in New York City. "What about Louise?" he asked quietly. Later that year, my parents were married.

Both my father and my mother were very interested in my education, which began at Sacred Heart High School in Waterbury. My father, a more or less self-taught experimental metallurgist, had an extensive technical library. He also worked very long hours and never took a vacation. Consequently, I usually spent my vacation periods with my Uncle Joe Anthony and my mother's sister Margaret and their family. This uncle was a zoologist who retained a child's curiosity for his work. I could hardly not have been influenced by a scholarly, hard-working father and a zoologist who saw fun in his work.

But the dominant impression of my youth was the sickness and death from leukemia of my only sibling, my sister, Judy. The ordeal brought me into contact with the nonfamily member who may have most influenced my future, Dr. Marcus Cox, the pathologist at Waterbury's St. Mary's Hospital. It was he who diagnosed Judy's leukemia in 1948.

Marcus Cox was the first true skeptic I ever met. He has also, in his ability to cut through folderol to the hard facts of things, come closer than anyone I know to those fictional prototypes of clear-headed rationalism (which I tend to associate with the British), such as Sir Arthur Conan Doyle's Sherlock Holmes and Umberto Eco's Brother William of Baskerville. Strongly built and square-jawed, he was given to brief conversation, always precisely if economically selecting his words. Out of character were his penchant for fast cars and a certain difficulty in refusing a Scotch and soda or a Manhattan with two cherries. A clinical pathologist by medical specialty, he had the attributes of a born scientist, ever curious, always seeking to find out why something happened the way it did, harshly critical of glib explanations. He had no children, which may have accounted for his quiet but intense interest in young people. He catalyzed in many students those latent talents that resonated with his own intense love of learning, understanding, classifying. Nothing made him happier than seeing a young person enter medicine under his wing.

At first my father would not accept his harsh diagnosis and prognosis for my sister and, needing to direct his anger and frustration outward, turned hostile toward him. But soon they became the closest of friends, a friendship that lasted until my father's death in 1983 and now continues through me.

Dr. Cox referred my sister to Harvard's Children's Hospital, whose director was the famous hematologist Dr. Sidney Farber. Now resigned to the gravity of my sister's illness, but still hopeful of a miracle, my parents held in awe Dr. Farber and his staff (I remember names of other doctors: Diamond, Daniels, Duffy) and were forever afterward grateful for their attempts to pull off the miracle.

Judy had been a happy, plump, healthy, and beautiful five-year-old. I was the eleven-year-old brother she adored, and no matter how hard I tried to lose her, she followed me everywhere. It was the summer of 1948. World War II was already three years in the past, having consumed nearly a quarter of my young lifetime. My mother's two brothers had both returned safely, one from an infantry unit that had been part of the African campaign and another from an Air Corps tour in England. Growth and change were everywhere in America, and our industrial-mill city was booming.

A hot day at the sea in Milford, Connecticut—Bayview Beach it was called. The great Joe DiMaggio returning one more time from an injury and marking the return with a home run off the almost equally great Bob Feller. Suddenly, chills and fever. Then better—only the "summer flu." The return of symptoms. The diagnosis, the panic. Some months earlier, Judy had put her arm through a glass pane in the door of my aunt's house, and the gash had run deep and ugly. Now that the disease was described as a blood disorder, my father first blamed my aunt, then a doctor who had treated Judy with a sulfa drug, and then Dr. Cox for making the cruel diagnosis.

I had not been told how seriously ill my sister was, so I was unhappy and complained during the long months while my parents were at Children's Hospital. Then, after living with one relative after another,

I was told that my sister was better and that I could visit her in Boston. The year was still 1948 and Judy was one of the first persons to be treated with antimetabolite chemotherapy, which uses agents that mimic a normal component of a metabolic pathway and so interfere with a chemical step in the path—in this case, with DNA synthesis. Because cancer cells need to make DNA before they multiply, and because most normal cells are not multiplying in a person who has cancer, the drug will kill most cancer cells more selectively than it kills normal cells.

Forty years later, the 1988 Nobel Prize in medicine would be given to George Hitchings and Gertrude Elion for this very work. Judy obtained one of the earliest clinical remissions from drug therapy ever; I think my parents thought she was cured. It was in an atmosphere of elation that my aunts and uncles told me I was going to Boston to see her. But in the few days it took us to get there, she relapsed. The jolly, plump, pretty sister I remembered was now emaciated, jaundiced, and covered with bruises. When she smiled I saw only the caked blood covering her teeth. This was the last time I would ever see Judy, and it remains the most powerful and frightening demon of my life.

Some weeks later (or was it months?—children count time differently), once again with my aunt, I was told to pray harder and longer for Judy one evening because she was worse. I did. The next day many of the members of our enormously extended family attended a mass together, and again I was told to pray hard. I did. After church, Uncle Joseph, the zoologist, asked me to accompany him and my grandfather in his car. The ride was silent. I sat next to my grandfather and watched him smoke his strange, foot-long pipe in silence. This in itself was not unusual. I had often watched him sitting and listening to opera on the radio or on records while he silently smoked this same pipe. Soon we arrived at my house. I had not been home in so long that I had not realized where we were going until we arrived. Suddenly, there at the back of the yard, I glimpsed my father and mother removing suitcases from their car, right below the basketball hoop my father had put up against the garage a few years back. I yelled out, "There's my mom and dad!" I started to run toward them, excited and happy, but Uncle Joseph grabbed my arm. "Robert," he said to me, "Judy is dead." It was March 1949, a few days before my twelfth birthday.

In my father's endless and profound grieving, his floating blame finally came to settle on himself. He had, after all, never been a believer, a faithful churchgoer. He would now become one. Conversely, as a child, I was a believer, as are all children. But the God I saw then was not to my liking. Though the church remained a part of my heritage, I never again thought of religion as a source of comfort or strength, or of prayer as an effective way of getting what I wanted or avoiding what I feared. My mother, too, was devastated by the loss of her only daughter, but as I look back now, it was her strength that held the family together through this period and the years to come.

In the midst of the horror of that experience there would also be some positive influences for me. Increasingly, I saw science as another kind of religion, certainly one that would yield more predictable results if one served it faithfully. Though I had seen firsthand the limitations of medicine, I had also seen another world, academic centers peopled by physicians who devoted their lives to trying to cure disease by perfecting their understanding of its mechanisms. Although I had no particular interest in science and avoided thoughts of medicine and sickness throughout much of my high school years, filling my days and nights instead with sports and social activities, I must not have let go of this early exposure to medicine altogether, for the books I remember reading through this time—Paul de Kruif's *Microbe Hunters,* Sinclair Lewis's *Arrowsmith,* Morton Thompson's *The Cry and the Covenant,* Francis Minot Rackemann's *The Inquisitive Physicians*—generally celebrated the satisfactions of a life spent in medical science.

Still, those who remember me from that time would remember sports as the center of my life. Basketball was a particular joy to me. I played day and night, summer and winter. The game allowed me to shed the overcoat of loneliness and confusion. When not practicing with the school team, I would put rugs on the frozen surface of our backyard and shine lights from the porch to allow me to play on into the night. I was devastated when a fractured sacrum one week before the start of the basketball season in my senior year in high school kept me off the squad.

Yet this, too, had its positive side—indeed, it reinforced in me the lifelong feeling that every negative, given enough time, produces a positive. At least I think this has been my consistent experience. Now free of the daily ritual of team practice, I had time to think about my future. My relationship with Marcus Cox became close, and I began to spend days with him in the hospital—even assisting in postmortem examinations. By the time I entered college I knew that I wanted to pursue medicine, going into biomedical research. Even more specifically, I knew that I would focus on the biology of blood cells. I had no sense—no conscious sense, at any rate—that I was moving toward the study of leukemia.

In Providence College I majored in biology, helped in a research project on cholesterol biosynthesis (looking back on it now, I realize that I had no real understanding of what I was doing), and became interested in the function of the thymus gland. At that time (the mid-1950s) no one had much of an idea about what this gland did. Following that era's crude approach to the study of organ function, I decided to remove the gland from healthy mice and observe what happened to them. In my last year of college, I obtained a few hundred mice and sent them to my mother to care for until I came home in the spring. Being a mother, she of course complied, and thereby made herself an accomplice. I performed the experiments; needless to say, no mouse survived my scalpel.

This incident confirmed three things I had already suspected about myself: (1) I would not make a first-rate surgeon; (2) I did not like to work with, and had no special gift with, live animals; and (3) I would not learn the function of the thymus gland. It also suggested that one should have well-developed technical skills before embarking on even the simplest of experiments. Though I left the thymus gland, it would not be forever. As a strange coincidence, the focus of my research team twenty years later would be on the thymus-derived T-cells.

Thomas Jefferson University School of Medicine in Philadelphia fortified my research interests, but not without some difficulties. I had to get through, for example, anatomy, microscopic (or histological) anatomy, and whole-organ-human (gross) anatomy. The experience with gross anatomy lab was indeed gross: dead bodies stinking of formaldehyde! It was said that you could tell when a student had come from gross anatomy lab—his nose remained defensively scrunched up for hours afterward. And how could one be sure of what we had to know? It was safe only to know everything—thus endless memorization.

These were difficult years for me, full of winter-gray, soul-lonely nights alone in my dark room, with its bare hanging bulb, only the bones I brought back from one of the laboratories to keep me company. Outside my window, neon lights directed near and would-be drunks to the next-door bar where, after further processing, they were somehow deposited in my doorway to be hopped over each evening. And of course, I have a clear memory of always being tired from the relentless cycle of study and work: six hours of studying on a short day, after a full day of classes and laboratory work. There were the dedicated and stimulating teachers who kept us going, but was this really what I had come for? Where were the real research laboratories—the accoutrements, the evidence of modern science?

Some medical schools have strong research programs, but the Jefferson University of 1959–1963, though strong in its clinical departments, and with some individual strengths in the basic science departments but not enough depth, provided few incentives to students to do research.* An exception was a blood research center, the Cardeza Center, the only pure research group at the medical school. It was my good fortune that Alan Erslev had just arrived at the center from Harvard Medical School. Handsome, animated, full of enthusiasm for his work, he was the lightning rod for my electrical charge. At Harvard, he and a few others had found evidence of a factor that induced growth of the progenitors of red blood cells leading to an increase in their number when needed. His discovery came from a study of rabbits that were made anemic by bleeding. The assumption (or maybe just Erslev's hope) was that the rabbits would deal with their anemia by regenerating their blood cells, releasing into the serum larger than normal amounts of some factor that might then be detected by Erslev and his team.†

The assumption was correct, and the growth factor they detected was the first discovered for any type of blood cell (though not the first of all growth factors). As we will see, many others were subsequently discovered for other types of cells, including one that would be central to my own work.

My interest in research also exposed me at an early point in my

*What Jefferson lacked, it often made up for in flexibility. Robert Wise, chairman of medicine, fostered a program in which I spent six months at Yale during my last year of medical school, just prior to internship. I was a "metabolic clerk" with daily involvement with Franklin Epstein, one of the greatest clinical investigators in the United States (then head of metabolic and kidney disease at Yale and now head of medicine at Harvard's Beth Israel Hospital). The profound care he took for the welfare of his patients and his influence in helping young people approach both a clinical research problem and (diagnostically and therapeutically) a person with a chemical-imbalance disease would stay with me forever.

†Similar but less complete (and largely forgotten) experiments were first done in the early years of this century by others in Europe.

career to the building-block structure of scientific inquiry. Everyone's work must stand on someone else's work or it does not stand well. Erslev's ideas had germinated from earlier work (in the 1940s and 1950s) done on the physiology of oxygen deprivation among high-altitude dwellers of the Andes and Himalaya mountains. For example, the South American physiologist Alberto Hurtado had found that the number of red blood cells correlated directly with the degree of oxygen deprivation found among these mountain dwellers.

Long afterward, the blood-cell growth factor found by Erslev's team was purified and the gene for it molecularly cloned. Now, by recombinant DNA technology, erythropoietin (or EPO, as it is commercially called) is available to treat some patients with anemia.*

Alan Erslev quickly became my inspiration, as well as providing me with an avenue into research. When I first came to work with him in 1961, as a summer research Fellow, he was interested in finding out exactly where—that is, on which of the many precursor cells of red blood cells, all of which are made in the bone marrow—this new growth factor exerted its effect. My job was simply to start the process by separating out from a bone marrow sample every type of bone marrow cell, from stem cell to the final precursor cells of the red-blood-cell lineage. I had about three months to do this.

I began enthusiastically. Using heat, I closed a piece of polystyrene tubing at one end, and I prepared albumin and sugar solutions of many different concentrations. I was hoping to create layers of different cell types by letting the cells of a rabbit's bone marrow settle in the solutions of different specific gravities. Aided by gentle centrifugation, the settling should make the larger, denser cells form a layer at one level and the smaller cells at another level, with infinite gradations in between. Then I would cut the tubing in several places, to conform to the height of the various cell layers that had been created, and isolate the different and relatively pure populations of cell types. The plan was not bad.

I did all my preparations carefully, and then went over them to make sure everything was right. After a while I realized that I was compulsively, obsessively, endlessly spending my time converting Erslev's laboratory into a glass menagerie of bottle after bottle of sugar

*Recombinant EPO has been put to unfortunate use as well. Since it raises the amount of red blood cells in normal individuals, at least temporarily, it can improve athletic performance in events requiring bursts of energy. Such use of EPO is called *blood doping*.

and albumin solutions that varied in their concentrations in minuscule, insignificant amounts. I could not get myself to conduct the actual experiment. My chief problem, though I could not face it at the time, was fear—fear of failure, fear of exposure, fear that I was not good enough, not smart enough, not trained enough technically. At the same time, Erslev also wanted me to help verify the strange notion, supported by early results of some laboratories, that the kidney was the source of this red-blood-cell growth factor. I never even started these experiments.

Finally, after a few months of this, Erslev communicated his displeasure to me one Friday. He did not scream, rant, or lecture; he just fumed. I felt embarrassed and dejected. The man whose approval I needed saw me as a failure. How could I ever make it in research?

Fortunately, over the weekend my anger and determination rescued me from my inaction, and on Monday morning I plunged in. I did a few of the tests, but they were not doing what I had hoped they would. I quickly concluded that I would not solve this problem this way and asked for a new assignment.

That was in 1961. Now, almost thirty years later, no one has yet succeeded in doing what I tried and failed to accomplish in Erslev's lab—namely, physical separation of all the cells of the bone marrow. I eventually learned that other scientists working on erythropoietin would demonstrate its source to be kidney cells, and even later (in the mid-1970s) that the major target cell for erythropoietin was one of the more primitive bone marrow red-blood-cell precursors. By then, of course, I would no longer be a medical student, nor would I be working with red blood cells. But I remained extremely interested in other growth factors, particularly those for certain white blood cells. Although I did not then relate one experience to the other, it may well be that what gave me the freedom to look elsewhere than precursor cells for the source of the white-blood-cell growth factor I was then studying was what I had learned so many years back about the source of erythropoietin. Today, the concept that cells of one type make biologically active molecules to regulate the production or metabolism and function of cells of another type is universally accepted and referred to as "cells talking to each other." Back then it was still a novel notion, except for hormones. But hormones were not known as growth factors, only as agents that alter cell metabolism.

The new assignment I asked Erslev for was a study of the response to oxygen deprivation in certain patients. I visited the lung division of the university hospital, where many former coal miners were being treated for some form of emphysema. Virtually all of them displayed some pathology related to oxygen deprivation. (I was reminded of the work by Hurtado and others on the mountain dwellers.) Even in the absence of any complicating factor, such as infection, these miners did not develop the appropriate red-blood-cell proliferative response to compensate for their lack of oxygen. I wondered whether the failure were related either to an insufficient production of erythropoietin or to an abnormality in its utilization by the red-blood-cell precursor. Though neither explanation proved to be correct, our research resulted in a detailed study of the underlying abnormality, which Erslev and I published the following year.

My first scientific publication! I was happy and proud. More important, I had begun to learn from Erslev a variety of laboratory skills, including how to approach a biological problem and, especially, to be suspicious and critical of one's own hypotheses and presumptions. Still more important, Erslev had forced me to dive into a problem, to commit myself fully, without fear of failure.

It was also through Erslev that I learned about the National Institutes of Health (NIH), and its postdoctoral fellowships. While a first-year medical student, in 1960, I paid a visit to its immense campus in Bethesda, joined by a cousin (the zoologist's younger son), two of my mother's sisters, and my fiancée, Mary Jane Hayes.

Mary Jane had also been born and raised in Waterbury, Connecticut, and we had both gone to the same high school. I was friends with her brother and knew of her through my high school years, but it was not until my first year of college that I began to date this pretty and kind woman. Four years later, the summer after my second year of medical school, we were married.

Now, back in 1960, as we walked around the NIH campus, I was overwhelmed by the sheer size of the place and by its imposing presence. I came away dazzled, unsure that I would ever be good enough to work there, but very much wanting to have the chance to do so.

After discussing my options with Erslev, however, we both thought it in my career interests to have a medical internship and perhaps a year or two of medical residency before entering NIH. My plan was to pursue an academic university position to do research, to teach, and to do some clinical work—the standard trilogy of the successful scientist-physician. As it turned out, NIH became the polishing ground for my final research training. But first, Erslev pointed me toward the University of Chicago, on the basis of its reputation as a major center of blood-cell biology.

Its head of medicine was Leon Jacobsen, a warm, intelligent, and nurturing man who had pioneered some aspects of radiation biology and much research into the red-blood-cell growth factor; he later became dean of the medical school. Of course, it was somewhat naive of me to think of research while I still had the responsibilities of a medical internship, but somehow, with generous help from the tall, lean, taciturn Jacobsen and his close associate, Cliff Gurney, I managed a small piece of research on the synthesis of the heme portion of the hemoglobin molecule. This would be the basis for my second publication.

I also came under the powerful influence of Rudi Schmid, who had just arrived at the University of Chicago from Harvard. As had Erslev, he had worked in Harvard's famous Thorndike Laboratories. Erslev had told me many tales about this extraordinary Swiss physician. Charismatic, handsome, brilliant—clinically as well as in the laboratory—he delivered crisp, clear, insightful lectures. Who could resist such a teacher? I began to spend many of my free moments under his generous tutelage and attended night meetings with a few of the young members of his staff. We arrogantly thought of ourselves as the avant-garde of the place. But soon I caught my first glimpses of intra-institutional politics—the bickering over policies, space, support. Schmid, with his multiple assets, but also strongly assertive, must have been seen by some as a threat, and within a few years Chicago unfortunately lost to San Francisco one of that era's great clinical scientists, administrators, and teachers.

As I look back on this period, and on the time I spent with these mentors, first with Erslev in Philadelphia (often at the expense of attendance at orthopedic or dermatology clinics) and then with Schmid and Jacobsen at Chicago, I realize that this was perhaps the hardest time of my life as far as work is concerned. Yet it is clear from this safe distance that it also provided me with some of my most exciting, gratifying, and happy moments, and many of my most vivid memories.

2

The National Institutes of Health and the Laboratory of Tumor Cell Biology

The National Institutes of Health evolved out of a little-known, but very important, predecessor public health institution, the Hygienic Laboratory. Virtually a one-person operation at the time of its founding in 1887, the laboratory was situated in the Marine Hospital on Staten Island in New York City, a government facility that provided medical care to merchant seamen.

In 1891, four years after its founding, the Hygienic Laboratory moved to Washington, D.C., and, thirteen years later, to better quarters a short distance away. The entire operation, still quite small, was devoted exclusively to practical public health issues. According to Dr. Bernice Eddy, one of the first women scientists to work there, the lab consisted of no more than two hundred people well into the 1920s and everyone knew everyone else.

In *The Health Century* (1987), Edward Shorter chronicles the emergence of NIH as the premier biomedical research facility in the world. By 1902, the Public Health Service (a government agency), through the Hygienic Laboratory, was helping local doctors deal with epidemic outbreaks of disease. In addition, it regulated the production of vaccines by the drug companies, many of European origin, that were just then opening their doors in America.

Early in this century, the U.S. public health system suffered from a glaring omission it did not share with the health systems of the major

European nations: it had no pure research facility where scientists and academic physicians might study the basic biological mechanisms of disease. Joseph Bloodgood of the Johns Hopkins Medical School, along with other academics, argued before Congress that the nation needed such a facility to complement the practical work of the Hygienic Laboratory.

In 1930, Congress responded. Renaming the Hygienic Laboratory the National Institute of Health, and making a firm government commitment to the funding of biomedical research, Congress was nonetheless careful to specify, as Shorter points out, that this research be "in the problems of the diseases of man," as opposed to pure biological inquiry.

According to Paul Starr, author of *The Social Transformation of American Medicine,* the decision to fund a more formal and far more substantial biomedical research institute was made at least in part in response to the successes of the Hygienic Laboratory in fighting epidemics. One of its most visible achievements had been the development of a rickettsial vaccine.

In the next few years, further changes strengthened the emerging NIH. First, and perhaps most important, academics could now come to work for NIH as civil service appointees; they did not have to become officials of the Public Health Service, which is one of the uniformed services. In 1937, when the National Cancer Institute, the first NIH specialized institute, was created, the 1930 legislation served to authorize the Public Health Service to fund outside private (mostly university-based) research for the first time, a significant innovation. NIH funds have, in principle, been available since that time to all researchers in the United States, and often to those outside the United States, willing to apply and have their projects peer-reviewed, a process NIH usually has conducted by experts in the given research topic drawn from U.S. universities.

Few nonscientists are aware of this situation, so that the inaccurate public perception is that the enormous NIH budget—today it is over $7 billion—goes strictly for NIH-based (or what is called "intramural") research, affording all those at NIH the luxury of money to spare. The total NIH budget supports almost all biomedical research conducted in the United States, including most of what is conducted at the university level. Surprisingly, it also supports much research in other countries. The groundbreaking work in molecular biology done by François Jacob and Jacques Monod at the Pasteur Institute (the French equivalent of NIH, though the Pasteur Institute has a commercial arm) was funded at least in part under an NIH grant.

In 1938, the NIH moved a third time, to its present home on what was then a forty-five acre estate in Bethesda, Maryland, that had recently been donated to the government. It now had more than enough room to grow. And grow it did. By the mid-1980s, Shorter notes, "there would be sixty-four buildings on three hundred acres." When a second specialized institute, the National Heart Institute, was established under NIH control in 1948, the National Institute of Health became the National Institute*s* of Health. By 1950 five more institutes were established; the annual budget had increased to $46.3 million. Three women—Mary Lasker, her sister Alice Fordyce, and Florence Mahoney—took up the cause of medical research and made their life's work the unceasing fight to pry increasingly larger sums of money out of the federal government to support ever larger NIH budgets for basic biomedical research. (Today the highest honor in biomedical research given within the United States is the Albert Lasker Prize, named in honor of her husband.)

Only once in its early history did the Institutes take a seriously false step. Shorter recounts that in 1954, during the height of the polio epidemic, the March of Dimes Foundation "ordered five drug companies to begin producing mass lots of [polio] vaccine, on the basis of a formula for inactivating the virus with formaldehyde, according to a procedure the polio researcher [Jonas] Salk himself had devised." Samples of the inactivated vaccine were then sent to the NIH Laboratory of Biologic Control, which was responsible for certifying that the vaccines were indeed inactivated and safe for use. The actual testing was done by Dr. Bernice Eddy, a scientist who had been with the Institute since the Hygienic Laboratory days (and who, along with the late Sarah Stewart, also of NIH, would be among the first to discover a DNA tumor virus—one known as polyoma virus, isolated from laboratory animals).

Using eighteen monkeys as test animals, Eddy and her staff began inoculating the animals with vaccine from each of the five drug companies. With one particular sample the test monkeys showed signs of paralysis—an indication that these lots of vaccine had not been properly inactivated. Although she reported her findings to the appropriate people at NIH, she heard nothing more about it. In fact, one of her superiors asked her whether she would like to have her own children inoculated, since there was not enough vaccine for every child in America. Not surprisingly, Eddy demurred. Shorter quotes her as saying, "They went ahead and released the vaccine anyway, a lot of it. The monkeys they just disregarded."

The only surprising thing was that the consequences weren't even more horrible than they actually were. Eighty children received active vaccine. Now harboring active virus, presumably through inoculation, they passed it on to approximately 120 additional people with whom they came in contact. By the time the error was discovered, three-quarters of the victims had been paralyzed and eleven had died.

A national furor followed. By the time it ended, the director of the microbiology institute had lost his post, and both the secretary of Health, Education and Welfare and the director of the NIH had resigned. The only shred of good to come out of this disaster was that James Shannon, one of the most important and forceful of NIH's directors, took over.

Starr talks about another phenomenon distinctly associated with the beginning of NIH as a research institution: the "remarkable degree" to which "control over research was ceded to the scientific community. The approval of grant applications as well as basic policy issues rested mostly [usually entirely] with panels of nongovernmental [non-NIH] scientists. The individual scientist, too, enjoyed autonomy within the constraints of professional competition." Not until the late 1980s would such freedom begin to be questioned or challenged.

By 1965, the year I joined NIH, Shannon was firmly ensconced and the Bethesda site looked very much like the beautiful and well-tended private estate it had once been. With the exception of one high-rise building, it consisted of no more than a smattering of low-rise laboratory buildings sprawled over its vast open spaces. Although Building 1, the first building completed and now the administrative building, had been renamed the James A. Shannon Building, all other buildings with only a few exceptions were numbered in the order of their completion, and remain named according to their numbers. Our lab is housed in Building 37, the thirty-seventh building to be completed.

One building stood out from the others, even back in 1965—Building 10, also known as the Clinical Center. Fourteen stories high, it is to my knowledge the world's largest building devoted to biomedical research. It is also the closest thing on campus to a hospital. Anyone coming to any institute of NIH for treatment will be housed in Building 10; patient services are also housed there. Only in its wings and its N (north) corridor are there basic research laboratories, again for all institutes.

I was filled with excitement and anticipation when I finally got to the National Cancer Institute of NIH, mostly because of support from Erslev, Jacobsen, and Schmid. It was a time when many young M.D.'s in this country idealized academic research, a state of mind that seems to have gone the way of the dinosaur. Most felt that spending time at NIH was virtually a prerequisite to academic medicine. Becoming what is called a clinical or a research associate was the thing to do. The war in Vietnam made an appointment even more desirable. Some who were not all that interested in a research career were still eager to enter NIH, as service in its public health officer program was accepted in lieu of a military obligation. The competition for spots became ferocious.

After a month or two the glamour and excitement wore off, and I realized that I was spending almost all my time not on research but in caring for very sick cancer patients. I soon learned that this would be the case for a full year. Worse, after six months I was assigned to the wards of the pediatric leukemia unit taking care of very ill children. We not only had to expose our fallibilities and inadequacies to the parents of our patients when treatment regimens did not work; we also had to be the ones to deliver the dreaded news to parents when a child died, which was often. Such experiences took their toll on each of us, on some more than others. One colleague began literally running away from the wards whenever a crisis with one of his patients seemed imminent, leaving it to whichever of his colleagues remained behind to deal with it. We hated him for it, but in some ways we understood. Another went to a nearby hotel and killed himself with morphine. This was my last direct involvement with clinical medicine.

Yet, in retrospect, there were some enormously positive components to that year. Cancer chemotherapy was just taking off for childhood leukemias and for adults with some forms of lymphoma. At first I was repelled by the aggressiveness of the chemical drug therapy, with all its toxic side effects, promoted by the then young Vincent DeVita (later to be the director of NCI) and by Drs. Emil (Tom) Frei, Paul Carbone, and, in particular, Emil (Jay) Freireich, all then of NCI, and others.* Soon, however, this work did lead to cures for Hodgkin's disease, some lymphomas, and some forms of childhood leukemia—including the very form my sister had died from almost twenty years earlier.

NCI clinical investigators were also among the pioneers in improving general supportive care. For example, the easy bleeding of a patient with platelet deficiency (one of the consequences of leukemic cells in the bone marrow) could not be properly handled at the time of my sister's illness. It was known even then that platelets, as a component of normal blood, could be partially restored by blood transfusions. In order to obtain a sufficient number of platelets, however, volumes of blood were needed in amounts that exceeded the capacity of a child's cardiovascular system, occasionally leading to heart failure. During the 1960s, researchers found a solution to this problem: the child would be given blood that had been specially processed to increase its platelet concentration, thereby reducing the total volume of blood needed. Jay Freireich, more than any other person, was responsible for these kinds of advances in supportive care.

Unfortunately, this kind of straightforward, practical research, too solution-oriented to pass as intellectual pursuit with some purists, did not enjoy a position of great respect in the academic community. In fact, I had been advised by more than one person to avoid associating myself with such work at all costs. I understood and partly identified with this sentiment coming from part of the academic community. After all, this research necessarily was chiefly empirical, not very intellectual, and often rather crude. But fairly soon I recognized that no area of research had a right to elevate itself above others, and that at times a tenacious persistence in the face of faint praise, or even outright criticism, is an important part of all but the most fortune-blessed success stories. When I came to NIH, I hoped my career would follow in the paths of those eminent predecessors in U.S. academic medicine of the 1950s and 1960s, pursuing the trilogy of a university-based medical professor of teaching, clinical care, and research. But I quickly learned that my inclinations were more toward

*Much of the credit for the successes of these years surely also goes to Dr. Gordon Zubrod, the director of the Cancer Treatment division. He remains for me a model of what a divisional director should be, supporting those under him while eliciting their best work from them. Later Zubrod would give his approval to the critical support I received in the beginning days of my own lab.

laboratory research and felt that in the future such research would become increasingly complex and require full-time attention. Indeed, this is exactly what happened. Few people in academic medicine today can do all three well. So there was no reason for me to hurry back to a university medical center. NIH was the place for me.

My first full-time research position began in July 1966, as an associate of Seymour (Sy) Perry, head of the medicine department. Coincidentally, Sy resembled Alan Erslev quite a bit, though not in his personality. Somewhat shy and gentle, he was not the type to criticize. He had come from Southern California, where he worked on blood-cell biology, and was one of the few researchers in the Division of Cancer Treatment to which I was attached who was not doing chemotherapy. His specialty was the study of the growth of white blood cells in various forms of leukemia.

There were four or five of us working with Sy on various problems in this general area, mostly on "cell kinetics"—the study of the correlation of the proliferation rate of different kinds of leukemic cells in a patient with disease progression, therapy, and so on. I chose a somewhat different path. Perry had a new collaboration with a biochemist, Ted Breitman, who belonged to NCI's Laboratory of Physiology and whose interest was in the biochemistry of the building blocks for DNA synthesis (the purine and pyrimidine nucleosides and nucleotides). Breitman had worked only with bacteria. The Cancer Institute had far more interest in blood cells than in bacteria, and more interest in white blood cells than in red because cancers (leukemias, lymphomas) occur in the white-blood-cells' primitive precursor cells present in the bone marrow, but only rarely in the bone marrow precursors of red blood cells. It was natural, then, for both Breitman and me to turn to the study of white blood cells, and this provided an opportunity for me to learn some biochemistry, a discipline in which I would need to feel comfortable in my future work. With Breitman I had my first real exposure to some general principles of biochemistry, as well as, more specifically, to enzymology—the study of enzymes, the proteins that catalyze all the chemical reactions of cells.

Sy was generous. He fostered an expansion of my work and gave me considerable freedom—his actions correlating, I assumed, with my increasing scientific maturity. Moreover, opportunities were growing. Sy was not only head of medicine but also the associate director for all the medical oncology branches. In the late 1960s he formed the Human Tumor Cell Biology Branch, which he at first headed up but which would soon provide me with the opportunity to have my own laboratory.

The building blocks of DNA are called nucleo*tides.* These molecules consist of a substance known as pyrimidine and purine bases, which are joined to a kind of sugar (called deoxyribose), which in turn is joined to one, two, or three phosphate groups. When the base is joined to the sugar without any phosphate, the molecule is called a nucleo*side.* We chiefly studied enzymes involved in the breakdown and synthesis of the nucleoside and nucleotide of thymine (base without sugar), the *T* in the four-letter alphabet of DNA. (The rest of the alphabet is *A* for adenine, *C* for cytosine, and *G* for guanine. As DNA precursors, they are each joined to a deoxyribose sugar.)

One difficulty we encountered was in the preparation of the enzyme under study. Most of the time this meant "purifying" the enzyme by separating one protein from most of the other components of the cell. This is hard work that requires spending a good part of the day in a room roughly the temperature of a refrigerator in order to avoid inactivation of the enzyme. I learned many skills of biochemistry in this atmosphere, including column chromatography, which usually involve packing a solid substance into a glass cylindrical tube (from a few inches to many feet in length) and pouring different solutions through it. For instance, one's sample, containing a mixture of constituent chemicals of the cell, is applied to the top of the column and allowed to flow through by having a solution continuously pour over it, sometimes under pressure. Different molecules interact with the solid and liquid components to a different extent and exit from the column accordingly, where they are separately collected. Proteins can be separated from each other and from many other molecules in the cell in this way; if a mixture is passed through, some components will come out faster than others.

More than one night I watched haplessly and helplessly in a cold room while one of my glass columns broke or ruptured at a joint. To hasten the flow of the solution, I sometimes put too much pressure in the system. I took refuge in Breitman's immense patience and tolerance with me, and in his recollection of a story about a dressing down given by a biochemistry department head to a young M.D. who years earlier had worked in the same room I was in. According to the story, he had caused a flood in the laboratory while washing his glassware and was told he would never make a biochemist. Some years later this young M.D. discovered the enzymes that make DNA (DNA polymerases, which I will discuss later), a discovery that earned him the Nobel Prize. His name is Arthur Kornberg.

There is something very gratifying about the "clean" results that can be obtained with enzyme work. In my first year with Breitman, I co-authored four biochemical papers, three of which were in the prestigious *Journal of Biological Chemistry.* On all four papers I was listed as the first author. Breitman gave me much-needed experience and new confidence. He also introduced me to a new term permeating the frontiers of biochemical enzyme research, *allosterism,* * and to its French creators at the Pasteur Institute in Paris, Jean-Pierre Changeaux, François Jacob, and Jacques Monod. I owed much to him and to Sy Perry. But as this period progressed I gradually began to feel that the problems in enzymology were not sufficiently biological for my interests, and that the results were not surprising enough or frequent enough. Furthermore, I did not see that I had any particular talent to bring to this field. I was playing to my weaknesses to some degree and not to what I perceived to be my assets, the broader biological problems—particularly those relevant to human disease.

I came to realize that I was interested in leukemias and their relatives, the lymphomas, and should not be afraid of involving myself directly in their study. What was their cause? Or were there many causes? Were environmental factors involved? Always? Never? Which would be the best of the many kinds of leukemias to focus on? Should I use animal systems or study human leukemias directly?

I thought that the leukemias and lymphomas that affected humans were probably an abnormality of white-blood-cell differentiation. Whatever their cause, the result was that the more primitive and proliferating precursor cells could not mature properly (and hence die), and so the cell population would gradually expand. Cell differentiation was known to involve programs for selectively making some new proteins and probably turning off the manufacturing of others. I wanted to know if the control of such events was abnormal in leukemia. Having been grounded in the fundamentals of biochemis-

**Allosterism* refers to a property of some enzymes to change dramatically their efficiency in catalyzing a chemical reaction under certain circumstances.

try, I now wanted to gain some firsthand experience in molecular biology—the evolving and exciting new area of research that seemed to be building the best tools with which to answer these kinds of questions. I was stimulated by the work of Marshal Nirenberg at NIH, who had just won the Nobel Prize for his research that led to the breaking of the genetic code. But by the time I was ready to make a move, Nirenberg was pursuing neurobiology, which was too far removed from my interests. One of his former associates, Sidney Pestka, however, was about to move to the National Cancer Institute from the Heart and Lung Institute of NIH, where Nirenberg was located. From his experience with Nirenberg, Pestka had become expert in transfer RNA molecules (tRNA) and in the overall mechanisms by which proteins are made. I decided to work with him.

In working with Sid Pestka on tRNA, I gained technical experience, awareness of molecular biology, and a new tendency to disbelieve almost everything. Whenever I ordered a compound from a pharmaceutical company, Sidney would ask me how I could be sure of what I had. For example, how could I know that tryptophan, a particular radioactive amino acid, was really tryptophan? I should first subject it to several tests, he instructed, including one to determine its movement under high-voltage electrophoresis. Once, after almost electrocuting myself, I asked Sidney whether perhaps we might be taking things too far. How do we really know our water is water? How do we really know our mothers are our mothers? Sidney was not amused.

Pestka remains to me a symbol of compulsive care in scientific experimentation, but he also showed critical intuitiveness. Going beyond his fastidiousness and critical tendencies, he would punctuate a philosophical discussion with startlingly practical insights. What I learned from him would stand me in good stead in the coming years when I would find myself in the position not of junior scientist working under someone else's direction but as the head of my own laboratory.

Through the late 1960s and early 1970s, although new buildings seemed to be going up everywhere and quickly, the NIH complex was only beginning to accommodate the needs of a newly expanding national scientific effort. Young scientists who had not proven themselves could easily find themselves and their staff in cramped or even makeshift quarters. I remember vividly my first "office," in 1970, located in Building 10. I was still part of Sy Perry's department and shared the office with two other postdoctoral Fellows. The office was no more than four feet by five feet, barely large enough for a desk. There was no way to open the door without hitting my chair (or my head), so the door was usually kept open. This created its own problem, for the office opened directly into the laboratory, and in those days I smoked. As Prem Sarin, who joined my lab in 1971, recently reminded me, even later with a slight improvement in office size, whenever somebody had to do an experiment in the lab, the first thing you had to do was "get Gallo out of his office," so that the smoke from his cigarette did not interfere with everyone else's experiments.

As these early years went by, conditions became no better for the young postdocs who now joined me. Marjorie Robert-Guroff, who came as a Fellow of the Leukemia Society in February 1971, recalled a summer spent in a wing of Building 10, where she "shared a single module of three hundred square feet with eight others, including one volunteer, a bored patient hospitalized in Building 10 who came over to help us out in the lab. In those days we could use all the help we could get."

Not surprisingly, there were no complaints when I announced that our group was moving, even though our new quarters would be off campus. Although the proximity to other working scientists gives working on campus a special intellectual excitement, our years at Pearl Street in downtown Bethesda are remembered fondly by all.

From the outside, Pearl Street was nothing special. As Marv Reitz, another postdoc in the lab, recalls: "The lab at Pearl Street looked like a roller rink from the outside. In fact, I believe it had originally been a skating rink." But Pearl Street proved to be a wonderful experience, and our years there were exceptionally productive.

Practically every NIH lab has contract relationships with private industry, which often allow a lab to use special technology without making enormous investments in equipment or personnel they will not use on a regular basis. Such contract relationships have long been part of government biomedical research.

Government oversight committees are never comfortable when NIH personnel work too closely with the people with whom they have contract relationships, however, so our move to Pearl Street, to space owned by Litton Bionetics, one of the firms with whom we had a contract relationship, may have raised eyebrows.

But the opportunity to work as one big group made up for our isolation. Another pleasure of Pearl Street was that if you needed to redesign your space—enlarge one lab, create office space for someone new—it could all be done within a reasonable period of time. This spoiled us. Government momentum is different. When we returned to campus and found that a requisition was needed just to get someone to change a light bulb, we had trouble readapting.

Even today, space limitations on campus require that administrative and epidemiological groups be regularly assigned off-campus facilities. For instance, for those at NCI, where I work, a special arrangement has been made with the Frederick Cancer Research Facility in Frederick, Maryland, to offer lab space to NCI workers in a facility used mostly by contract people. This is the only facility I am aware of where it was understood from the beginning that government employees would work side by side with private industry employees. Though not uncommon in some other government departments and though useful and needed, close contacts of this kind create concern among Health Department officials that such arrangements make easy targets for those looking for the appearance of wrongdoing or conflict of interest.

Today, without question, the NIH is the largest and most productive government-funded laboratory in the world. In 1987, during the centenary year of U.S. government support for biomedical research, "scientists and policymakers alike," according to the newspaper *The Scientist,* "hailed the NIH as the 'crown jewel' of the U.S. government's biomedical research enterprise." More recently, *The Scientist* noted that "NIH employs no less than 10 percent of the world's 100 most-cited biomedical scientists of the last decade." At the same time, others, particularly spokespersons for groups that feel that government-supported scientific research has become unresponsive to public crises, and those who have a particular bone to pick—animal rights activists, for example—have made NIH researchers the target of some harsh criticism.

Within NIH, the complaints take a different form. Never before have staff members—from the secretarial and administrative people to lab technicians, postdocs, senior investigators, all the way up to branch chiefs and institute heads—had to deal with so much outside-imposed regulation regarding documentation and ethics. The decision-making process at every level is subject to a kind of procedurally induced skeptical scrutiny few would welcome.

The sense of scientific freedom and individuality that marked the beginnings of the NIH was carried forward for me in the creation and organization of new laboratories within each institute and in the informal relations that sometimes established themselves from lab to lab, institute to institute. Within the National Cancer Institute, for example, some labs cooperate with other labs; others operate totally independently. Cooperation might be between lab chiefs or between postdocs. As important, even the organization of labs differed significantly from one to another, particularly in the early 1970s when the Tumor Cell Biology Lab was first starting up.

I was fortunate in coming to the NCI when I did. In 1971, the National Cancer Act was passed by the Nixon administration. Funding for cancer research was made a national priority, ensuring the continued availability of on-campus positions and research facilities. But other provisions of the act were even more important to the NCI as an institution, because they strengthened its position vis-à-vis the NIH, giving it a degree of freedom and autonomy that would become increasingly important in later years.*

One of the act's key provisions called for the NCI budget to be routed directly to the president, thereby bypassing the normal channels of review, which included the office of the director of the NIH as well as the Office of Management and Budget (OMB). But the only president to employ a bypass NCI budget was President Nixon; later administrations reverted to the earlier arrangements, routing the NCI budget through customary channels, though the possibility each year of a direct submission by NCI remains a presidential option under the Cancer Act.

Other provisions of the act have fared better. The NCI Advisory Board created by the act gives the NCI director the right to appoint advisers without consultation with the office of the NIH director. Possibly more important, the President's Panel, another creation of the act, continues to give the NCI direct access to the White House—an increasingly important provision during highly political periods. Not surprisingly, the two decades that followed passage of the act have been very productive years for NCI, and for cancer research in

*Although my lab was then officially part of the Cancer Treatment Division of the National Cancer Institute, John Moloney, who was then running the Virus Cancer Program in the Division of Cancer Cause and Prevention, passed funds from his program to our laboratory, hoping our work might help his.

general. In fact, I believe it fair to say that in many respects the enhanced funding and status endowed by the act were a significant factor in the propulsion of molecular biology, and more particularly of what might be called molecular medicine. But at the same time, the special status of NCI has sometimes put it in conflict with Building 1, as the office of the NIH director is sometimes referred to on campus.

Clearly the special status of NCI (and, to some extent, also now the Heart and Lung Institute) is bound to be a double-edged sword. The alternative, however, would likely have been worse. Between the 1950s and the 1970s, NCI, once the only institute, became but one of over a dozen institutes at NIH. Equality with the other institutes, while fair on the surface, wasn't always appropriate, because it often meant more delays, reduced funding, and increased bureaucracy in an area where the country had clearly called for expedited action. The pressing calls for breakthroughs in cancer research were too clamorous to allow this research to continue to be carried on in a business-as-usual manner.

With the exception of my first year, when I was doing clinical work in Building 10, I was funded as part of Sy Perry's Branch (or Department) during my early years at NCI. Officially, I remained part of his branch even as I moved in and out of a variety of labs attached to other divisions in NCI run by other senior scientists, including Sid Pestka and Ted Breitman. Back then there were three intramural NCI divisions—Biology and Diagnosis, Cause and Prevention, and Treatment—and an extramural branch that dealt with the administration of grants and with contract arrangements. Perry was the Medicine Branch chief in Treatment. But in his position as the associate director for medical oncology, he helped create a new department—the Human Tumor Cell Biology Branch. In 1972, he appointed me head of this department. I immediately changed its name to the Laboratory of Tumor Cell Biology. Today at NCI, much the same organization exists, except that Cancer Causes is now its own division and it is called Cancer Etiology. My lab was appropriately moved to that part of NCI in the mid-1980s.

It was not uncommon back then for a major lab to be organized around a single scientific problem—the three-dimensional structure of proteins, for example. The lab might consist of all senior scientists, working pretty much on their own with their own young trainees (postdoctoral Fellows). Other labs might be organized around a single facet of technology—electron microscopy or tissue culture or primate biology. Although the institutes attract both Ph.D.'s and M.D.'s, those labs doing more basic science—molecular biology or biochemistry, for example—tend to attract more Ph.D.'s than M.D.'s. The labs that attract more M.D.'s have tended to be clinical, that is, working directly with patients in experimental programs or, if working on basic science, treating a problem with immediate clinical relevance. But this is far from universally true. Arthur Kornberg, the discoverer of the first known DNA polymerase enzyme, is an M.D. who began his career at NIH. Conversely, a great deal of the medically practical work leading to the development of anticancer drugs at NCI is done by Ph.D.'s.

In the early 1970s, I had only a general idea about how to shape a lab. Although I was an M.D., I knew I did not want to work with patients in clinical medicine. But I also knew from my years in pure research that I did not want to work solely on biological problems, or to work on them completely apart from their medical implications. I was and remain to this day interested in basic biological phenomena, invariably with the anticipation that an understanding of them will somehow help us understand abnormal biology—that is, disease.

I ended up creating a research laboratory centered on a medical problem. My lab would, I hoped, employ the tools of molecular biology and biochemistry to discover and understand the basic mechanisms underlying the development of certain human diseases, even if we could not find their cause. The particular medical diseases I wanted to focus on involved abnormalities of the blood cells, in particular, growth of the white blood cells—leukemias, lymphomas—and deficiencies of white blood cells (their failure to grow).

From the beginning, the Laboratory of Tumor Cell Biology (now usually simply called LTCB) attracted a hodgepodge of clinical people—Ph.D.'s next to all kinds of M.D.'s, those interested in basic research, and, occasionally, epidemiologists. As with many other NIH labs, ours also attracted a truly international group of senior scientists and postdoctoral Fellows. Facilitating their decision to work here was the task of the late Congressman John Fogarty of Rhode Island who, along with a few other representatives and senators, worked to pass legislation to promote the growth of NIH and to make it more than a national government showpiece of biomedical science. Fogarty, in particular, wanted to extend its influence throughout the world.

Fogarty *scholars* are senior scientists, almost always from abroad, whose year at NIH is funded by the U.S. government. Mostly they have teaching and writing responsibilities and are often asked to participate in the intellectual life of NIH. Fogarty *Fellows* are young, foreign postdoctoral students who work with us in the laboratory, usually for one to four years, also funded by NIH. Many other foreign postdoctoral Fellows come to our lab with their salaries paid by their own country. Because we do not have graduate students, as we would in a university setting, we rely more than do university labs on postdoctoral Fellows. Except for those few who bring their salaries with them, this inflates the budget of an NIH lab compared with that of a university lab, because graduate students usually work for little or no pay. Over the years, these foreign postdoctorals have filled many of our positions, particularly in recent years when better salaries lure American students straight into industry. I currently have in my own laboratory senior scientists and/or (usually) postdocs, and even technicians, from China, Taiwan, Iran, Japan, Hungary, India, Pakistan, Sweden, Israel, Nigeria, Italy, France, Germany, the Philippines, Belgium, and Australia, most through the Fogarty program. In past years, I also have had people from Bulgaria, Czechoslovakia, the Soviet Union, Spain, Venezuela, Colombia, Peru, Scotland, England, Switzerland, Canada, Jamaica, and Trinidad, among other countries.

Because the lab started up at a time when the field of molecular biology was rapidly evolving, it soon became apparent that I would have to make certain decisions about how to run a lab that would mean I would no longer physically perform the experiments.

The term used among biomedical scientists to describe hands-on scientific experimentation is *benchwork.* Doing benchwork can be one of the most grueling and exhausting experiences, yielding few or no results. Or it can be grueling but exhilarating, yielding a catalogue of information about a particular scientific question. Occasionally it is dramatically exciting, particularly when hard work leads to a real breakthrough. But in all three situations, it is at times the most taxing and tedious work that science calls for.

Every new scientist does benchwork. I learned a great deal from the years I spent at it, years I remember vividly not only because of what I learned but also because of the long and lonely hours. As I discussed earlier, much of my time was spent in a cold room, the equivalent of a large refrigerator, purifying enzymes and other components of the cell that do not survive for long after being released from the cell into the temperatures of a comfortable human environment. In the early 1970s, the technology of molecular biology was changing so rapidly that anyone who did not make a special effort to keep abreast of a substantial number of these newly developed techniques would, in a few years, fall hopelessly behind even the students at the major universities. Gene sequencing, for example, was just coming into its own, as were major advances in immunology, such as monoclonal antibodies.*

No senior worker broadly interested in biology could choose to ignore these new techniques without paying the high price of being condemned to work in one limited area of research, severely constraining the range of biological questions that could be addressed. But if one had an interest at that time, as I had, in broad biological questions, it would be almost impossible to run the lab and be an expert in all those new techniques.

It did not take long for me to realize that the only way we could stay on top of the new technology coming from diverse fronts would be for me to form the lab around several nuclei of young postdocs—some under my direct supervision, others supervised by the more senior members of my group. We looked for postdoctoral Fellows who showed an eagerness to learn new techniques. Whenever we read in the scientific literature about some new technique in another lab, we would send one of them there to learn how it was performed. That person would then come back and teach the rest of us. This early decentralization of expertise is possibly the most important logistical decision I made for the success of our lab over the years.

In the area of molecular biology, David Gillespie and, in time, Marv Reitz would be the first experts in the laboratory to use molecular hybridization as a tool to ask genetic questions. Flossie Wong-Staal, a molecular biologist, joined the laboratory as a postdoctoral Fellow and stayed on to evolve into one of the major players in my group.

**Sequencing* a gene means determining the exact type and order of the four nucleotide building blocks—for example, ATCCGT . . . —that make up a gene: that is, the portion of a DNA strand that can code for a specific protein. *Monoclonal antibodies* are those that are made by one *clone* of B lymphocytes. They are highly specific, usually reacting with only one precise segment or configuration of a protein.

Because of her insight and leadership qualities, she gradually assumed a supervisory role; in that capacity, she worked with some outstanding postdoctoral Fellows. Riccardo Dalla Favera and Edward Gelman were among the first in our building to master the arts of recombinant DNA technology, which includes the cloning of genes. Steven Josephs spent time developing gene-sequencing technology. Mary Harper brought a special variation of the technique of molecular hybridization to our group. Called *in situ* hybridization, it can help determine the location of genes on human chromosomes as well as the activity of certain genes. It was an active and exciting period in which we, individually and as a group, continually conquered new techniques and incorporated them into the lab's growing technical skills. More important, we were able to use these techniques to approach fascinating biological questions at the molecular level.

And there was learning of another kind going on. Marjorie Robert-Guroff had come to the lab in 1971 with a Ph.D. in biology, to do work on reverse transcriptase. To my everlasting shame, I thought it important during the job interview to ask her whether she considered herself "a woman and mother, or a scientist." To her credit, she asked in reply, "Isn't it possible to be both?" If, because of my inane question, Marjorie had rejected me as a potential boss, and had not made her contributions to the lab, who knows how long it might have taken me to realize that no lab today can be competitive without taking advantage of the enormous pool of talent represented increasingly by women.

Within a few years I also realized that if I wanted to concentrate fully on scientific work and administration, I would need to delegate responsibility for personnel matters, day-to-day financial matters, and overseeing contract decisions to a senior administrative head. Fortunately, Prem Sarin brought just these skills to the lab in the early 1970s.

For the next decade, my colleagues and I at the Laboratory of Tumor Cell Biology would participate in a historic search—for the first disease-causing retroviruses capable of infecting humans. But before I go on to that work, let me give some necessary background into the world of microbes, particularly the viruses with which we are concerned.

3

Microscopic Intruders

In the continuous and dynamic relationship microbes have had with humans and other animals, some have actually been helpful and many have been innocuous. Others have been harmful under certain conditions, and a few fatal even under the best circumstances. Microbes also vary greatly in size, structure, and biology. Many bear structural and functional similarities to cells, which can be detected under an ordinary microscope: they contain a nucleus and cytoplasm; they carry out metabolic functions, including the synthesis of some molecules and the degradation and disposal of others; and they may reproduce on their own outside a cell.

Protozoan parasites—for example, the various harmless and harmful amoebae—fall into this class. These largest of microbes are in the range of 10^4 nanometers (nm). (One inch is about 25 million times bigger than one nanometer.) So do fungi, which are also unicellular, nucleated microorganisms. The important break occurs with bacteria, which, though maintaining reproductive ability independent of an animal cell, show major differences in size and complexity from protozoa and fungi. Bacteria are smaller than protozoa and fungi, and much smaller than human cells. They are also less compartmentalized. For instance, they have less specialized functions and lack a nucleus.

In this descending order of microbes' size and complexity, rickettsias and chlamydias are next in line. Sometimes they cause disease:

for instance, Rocky Mountain spotted fever is caused by a rickettsia, and some venereal diseases are caused by chlamydia. Like bacteria, these are unicellular microorganisms that lack a nucleus and reproduce themselves by the process called binary fission. But they differ from bacteria in a fundamental way, one that brings them closer in some respects to viruses: they are almost always obligate parasites (that is, they absolutely require a cell in order to reproduce themselves). The metabolic functions they are able to perform independent of cells are not sufficient to sustain them, and unlike protozoa, fungi, and bacteria, the rickettsias and chlamydias cannot reproduce themselves apart from the cell.

Still another microbe with properties in between those of a living cell and those of a virus are the mycoplasmas. These microbes are also called pleuro-pneumonia-like organisms, or PPLO. They are the smallest known living cells (approaching the tiny size of viruses) but differ from viruses and from chlamydias and rickettsias in being able to reproduce outside a cell. Thus, like the bacteria and the fungi, they can be cultivated in a cell-free media.

There are many known human mycoplasmas. One causes pneumonia but the rest apparently cause no harm. Their biggest current claim to fame is their frequent, annoying presence as laboratory contaminants of virus preparations and cell cultures. Because they lack the rigid cell wall of a bacterium, they can assume different odd shapes, often resulting in their being confused with a virus. Scientists have sometimes been embarrassed by reporting their role in one or another disease, only to learn later that mycoplasmas were accidental bystanders.

Viruses, which are central to my story, are the next class of microbes, one that has been described as bridging the inanimate and the animate. Viruses are smaller than the microbes I have already mentioned. Robert Shapiro, a DNA chemist, geneticist, and the author of *Origins,* has helpfully depicted this relation on a 10^6 scale, where a bacterium is the size of a man and a virus is about the size of an arm. This general guide is not always accurate, however, as the largest viruses (about 300 nm) begin to approach the size of the smallest bacteria (about 500 nm). Other researchers have drawn even more dramatic comparisons, for example, that 1 billion billion of the smaller virus particles would be needed to fill a Ping-Pong ball. If this ratio is put into a time comparison, then a virus is to a Ping-Pong ball as one-quarter of a second is to the age of the earth. Special magnification by an electron microscope (EM) is needed to see viruses (usually magnified at least fifty thousand times). Many viruses approximate 100 nm, almost the exact diameter of the AIDS virus. The forms of viruses also vary greatly, from rod-like structures and spherical particles to more interesting and complex geometric forms.

But it is not so much the size or form of viruses that is of interest to us; it is their biological properties. Like rickettsia and chlamydia, viruses are obligate cellular parasites. But they are quite efficient at this and do what they have to do without many of the tools that rickettsia and chlamydia must carry. Outside a cell a virus is a piece of nucleic acid protected by some protein covers, a chemical entity simply following the rules of physical chemistry. In several cases viruses have even been crystallized. This allows us to learn about their real form. Sometimes they have an outer additional membranous envelope, which helps in their transmission from cell to cell.

Once inside a cell, specific viruses will, in time, exploit the machinery of certain cells to their own reproductive advantage but usually to the detriment of the healthy functioning of the cell. This parasitism makes use of subtle, specific, and ingenious means that have been evolved to allow specific viruses to enter and then replicate within specific target cells. These means vary from one virus type to another. Viruses are the ultimate *agents provocateurs* of biology, for they appear to be welcomed into the trusting arms of the cell.

The targets of the virus, those specific cells they infect (scientists call this phenomenon their *tropism*), vary considerably from virus type to virus type. But *all* of the plant and animal kingdoms are homes for one virus or another, and many virus families span the vertebrates and invertebrates, though one particular virus in a family is usually very limited in the range of species it can infect. The discoveries of viruses were made by scientists studying infections of plants. Other scientists studying bacteria would soon find viruses that specifically infect bacteria. Called *bacteriophages,* these bacteria viruses were discovered in 1916–17 by D'Herrelle in France and Twort in England and soon provided the basic laboratory systems scientists used to begin the study of modern molecular biology—an era spawned by Max Delbrück and Salvatore Luria in the United States.

At first viruses were classified only according to their effects on animals. (Pasteur began this with his study of rabies in mice and in dogs, though he did not know he was dealing with a virus.) With the development of electron microscopy in the 1940s, classification could then also be by size, shape, or even on the basis of finer structural details of the virus. By the middle of this century, laboratory systems for studying viruses were developed. Scientists no longer had to induce disease in animals to observe the biological effects of virus infection. To distinguish work performed using laboratory systems from work with live animals, scientists use the term *in vitro* for the former and *in vivo* for the latter.

The development of *in vitro* systems greatly facilitated progress in this field, enabling scientists to make better quantitative estimates of the amount of virus in a sample, to determine the target cells of a particular virus, and to see whether the virus produced a cytopathic effect (killing a target cell or otherwise causing the cell to alter grossly its form and function).

In many of the early laboratory systems, viruses were grown in chick embryos, which, strictly speaking, were still *in vivo.* More recently, many viruses have been grown in animal or human cell cultures—that is, in cells taken from the blood or tissue of infected individuals. Without help from the scientist, most such cell cultures can be kept alive for only brief periods of time, usually days. Hence, they are called short-term (or sometimes just primary-cell) cultures.*

Occasionally, however, for reasons we do not fully understand, and with no further help from the scientist except periodic nourishment, a short-term culture will take off and cells that would normally die in a few days acquire the ability to grow indefinitely. When this happens, the cell is said to have *immortalized* and the resulting culture is called a *cell line.*

It didn't take long for virologists (and other scientists) to realize the potential of the cell line as a research tool. If a virus could be transmitted to a cell line (without completely destroying the cell line in the process, which sometimes happens), many of the problems associated with short-term culturing would be solved. The culture itself would consist of a single cell type continuously multiplying. The virus could be produced in relatively pure form. And most important, the scientist

*Although techniques for tissue culture were a development of the 1930s, the techniques have been continually refined up to the present. A big advance came in the 1950s when John Enders grew the polio virus in cell culture. The 1950s and 1960s became one of the great periods of medical virology because of these cell-culture advances.

would be assured of large enough quantities of virus, obtainable repeatedly, virtually on demand, to study the reproductive (replication) cycle of the virus and its chemical makeup.

One important clinical consequence of being able to determine the chemical makeup of a virus is that scientists can identify the core and envelope proteins specific to that virus. With this information clinicians can then survey the general population for infection by a particular virus by testing for the presence of antibodies to these viral proteins in the blood of each individual. With cell-line culturing of a virus, sufficient amounts of the virus can be obtained for this and other studies.*

With a substantial and reliable quantity of virus to work with, scientists could also inject into an animal either the whole virus or a harmless component of it in an attempt to provoke the production of antibodies against the virus. The antibodies made by the animal (called *hyperimmune sera*) would always be specific to the class of virus with which the animal had been injected. This hyperimmune sera helped scientists further classify viruses. Those of the same type were soon found to show precise immunological cross-reactivity to the same antibody. Those less related to the given virus would show weaker cross-reactivity, while those more distantly related usually show no cross-reactivity. Later, in the 1970s, the monoclonal antibody technique of Georg Kohler of Germany and César Milstein of England produced highly specific antibodies that reacted only with a certain site of a protein.

Around the time the first laboratory systems were being developed, scientists began to be able to analyze many viruses chemically, leading to their broad subclassification according to their nucleic acid (genetic) content: deoxyribonucleic acid (DNA) viruses and ribonucleic acid (RNA) viruses (the nucleic acid may be single- or double-stranded). They were also getting new information on what might be called the life cycle of many viruses. Some viruses burst from the host cell, destroying it in the process. Others "pinch off" or "bud" from

*A cell line remains essentially the same over a long period of time; its protein components do not vary significantly. This is important when testing sera for antibodies to a specific virus in order to determine whether a person has been infected or exposed, because when a virus is grown only in primary cells, those cells will soon die. New cells from new donors must always be used, and they will have some protein differences from the earlier cells. This can interfere with testing for the antibodies to the virus. In this context, for example, it can be safely said that cell-line culturing of the AIDS virus was the breakthrough needed before a blood test for the AIDS virus could be successfully achieved.

the cell membrane, giving them an added outer envelope, which includes fatty components from the cell membrane.

Sometime between 1950 and 1960, then, viruses could be described by the diseases they caused, their nucleic acid content, their structural appearance, their chemical composition, their relation to one another through immunological cross-reactions, and the presence or absence of an outer envelope. These advances, combined with the advances in molecular biology techniques that have occurred since the 1960s, also made it possible to understand the genes of viruses, the functions of the proteins coded for by these genes, and the replication cycles of viruses.

The amount of genetic information a virus has (size of genome and number of genes) varies greatly. Smaller viruses contain few genes but some larger viruses, like some herpes viruses, can contain more than one hundred.

Unlike animal cells and most other microorganisms, a virus does not reproduce by growth in size followed by division. Instead, it enters a cell and, with the help of the cell itself, synthesizes what it needs and reassembles these elements into new virus particles. There are two main elements that must be synthesized for viral replication to go forward: (1) the proteins needed to give the virus particle its coat and structure, as well as to facilitate other biochemical processes, must be made, and (2) the viral genome itself must be copied. Although the virus is able to utilize certain factors and enzymes of the cell to accomplish these tasks, its ability to make use of host-cell machinery is not unlimited. At a minimum, all viruses must bring to the host cell those genes that correctly code for these proteins, as well as the know-how to duplicate their own genome, which will be in the form of either viral RNA or DNA.

It is a simplification, though for most purposes a convenient one, to say that all viruses must carry within their genomes the genes to make viral proteins. More precisely, what these genes must have or must be able to code for is the correct messenger RNA for those proteins. It is the cell's own machinery that will "translate" these mRNAs into viral proteins. This means that, with DNA viruses (such as various herpes, hepatitis, smallpox, and papilloma), the virus must be able to make its way into the nucleus of the cell; only there can the cell make mRNA out of DNA (with an enzyme in the cell nucleus called RNA polymerase). But some RNA viruses can make proteins without getting into the nucleus of the cell, because some of the RNA carried in the virus particle can act as mRNA as soon as it enters the cytoplasm of the cell, thereby directing the production of the viral proteins.

But there is a special problem for animal and human RNA viruses in carrying out their second task—reproducing the RNA genome—because, unlike plants, animal cells cannot make RNA from RNA (animal and human cells do not contain the proper enzyme). To reproduce its genetic information in animal cells, an RNA virus must carry one or more special enzyme(s) to create this capability. Some DNA viruses also carry their own enzyme for copying their DNA, although it is possible for some DNA viruses to use the cell's enzyme (a cell DNA polymerase) to duplicate its genome. So we see that different viruses have evolved amazingly varied and clever strategies to solve their problems.

Of all RNA viruses, it is a special kind—a virus form or type that will turn out to play a key role in some cancers and in AIDS—that contains a unique enzyme (reverse transcriptase) to allow these RNA viruses to have their genetic information converted into DNA (RNA → DNA). We say these RNA viruses have a DNA intermediate form in their replication cycle because they go from RNA to DNA and then back again to RNA forms (both *messenger*, to participate in making the viral proteins, and *genomic*, which soon will be assembled with the viral proteins in the final stages of virus formation). These special viruses are called retroviruses because of their ability to work backward from RNA to DNA instead of through what were previously known as the only flow paths of genetic information, namely DNA → DNA or DNA → RNA.

Though the precise tactics vary, the general steps of infection of a cell by different viruses are similar and can be summarized as follows. The first step is *attachment* of the virus to the surface membrane of its target cell. For an enveloped virus, this is usually accomplished by a specific interaction (binding) of a portion of the protein of the viral envelope and some protein molecule on the cell surface. After this interaction takes place, a poorly understood process occurs that allows some viral components to penetrate the cell, and in this process the fatty component of the viral envelope will fuse with a fatty layer of the cell membrane. The location on the cell membrane where this can occur is called the *receptor site* for the virus. Obviously this receptor site must normally perform functions helpful to the life of the cell.

Alternatively, as in the case of unenveloped viruses and sometimes for enveloped viruses as well, *penetration* may involve a process called *endocytosis,* in which the particle is more or less engulfed by an invagination of the cell membrane, which then closes in on the particle.

The fate of the swallowed-up particle varies among different viruses, but the synthesis of viral proteins and the reproduction of the viral genome are essential for the survival of the virus. As noted, to make viral proteins requires a viral RNA in a form (messenger RNA) and location (cytoplasm) that can be "read" or "translated." The order of the nucleotides of the messenger RNA determines the transfer RNA (tRNA) called for, which will provide the correct order of amino acids, the building blocks of a protein. For some RNA viruses, acting as mRNA is rather direct and simple. For others, the RNA of the virus must first be modified.

The final stages in the replication of any virus involve the *assembly* of viral nucleic acids and proteins. Those that fully assemble and thereby acquire infectivity while still inside the cell will be *released* from the cell only when the cell bursts, thereby simultaneously releasing the offspring virus particles and causing the cell's own death. The bursting is believed to be brought on by the harmful effects on the cell of the proteins of these kinds of viruses, but our understanding of the phenomenon is incomplete.

Enveloped viruses do not generally complete their assembly within the cell, and their exit from the cell occurs in a more economical and aesthetic fashion. They combine their final stages of assembly with their exit from the infected cell through a process of budding into the extracellular space.

In this second category are viruses having a range of effects on cell structure and function, only occasionally killing the cell. The AIDS retrovirus is an exception to this general rule: it is enveloped and buds from the cell membrane, but its formation often leads to the death of some of its major target cells.

As it is released from the cell, having completed its replication cycle, the virus is now free to infect another target cell. Some of the viral particles will succeed. Many will not because they will be defective, that is, not all will contain the full necessary amount of the viral genetic information to produce a complete and infectious virus particle. This happens because of the tendencies for viruses to lose some of, or to modify, their genetic content. Others will not be successful because they do not reach a target cell before being degraded in the body or "bound up" by the attack of our immune system. When large amounts of virus are formed and released at a certain time, the viruses can be found "free" (that is, not in a cell) in the blood plasma. We call this a viremia. Usually, it is very transient and associated with acute clinical symptoms, such as fever and headache. Other viruses are not produced at any time in sufficient amounts to give rise to a viremia. In fact, it seems some are transmitted from one individual to another only or chiefly by cells, the virus entering a new cell as two cells come in contact with each other.

We do not know the origin of viruses, but happily the old arguments that tried to settle whether they are living or nonliving must have grown tedious because we no longer hear them. Nor need we consider whether viruses represent an evolutionary link between nonliving and living matter. Because they require a living cell to survive and to replicate, they could not have existed as a life form before the existence of a living cell.

Some scientists have speculated that in a weird case of reverse evolution, viruses are descended from more complex parasites, adapting by shedding their ingredients to the barest form capable of survival as the thieves in the night they have become. Others have suggested the opposite—that viruses derive from even simpler forms (called *viroids,* to be discussed shortly), acquiring greater complexity. Many students of retroviruses, myself included, believe that viruses are escaped cell genes with a capability for diversification related to their special method of replication. It is possible that their "escape" has served useful functions in evolution, possibly by moving genes from animal to animal and even species to species. If you like this theory, retroviruses are your example because of their wide distribution among animal cells and their extraordinary ability to integrate their genes into the cell chromosomal DNA, as I will later discuss.

It would seem that as pieces of nucleic acid genetic material protected by a protein coat of one kind or another, viruses must be the simplest of all possible infectious agents. A pure protein can be neither infectious nor pathogenic since it has no mechanism for specifying itself—no genes—and therefore, almost by definition, no ability to replicate. On the other hand, nucleic acids alone cannot be infectious, because as potent as they are within the nucleus of a cell, they cannot negotiate the attachment to and penetration of a cell. Further, an unprotected nucleic acid would be quickly degraded.

These statements were surely made with justification before 1970. But today we know they are wrong. Quite astonishingly, some naked genes, that is, a nucleic acid without a protein coat (RNA forms), have been shown to cause infectious disease in plants. They now are the most extreme form of intracellular parasites. These RNA molecules are very small, with little potential to code even for a small protein. Discovered by T. O. Diener in Beltsville, Maryland, and called *viroids,* these RNA molecules can exist in some plants in a nonpathogenic state. When present in the wrong plant cell at the wrong time, however, they can lead to disease. We—or at least I—do not understand why. Most viroid diseases appear to be new. Apparently, they are transmitted chiefly by mechanical manipulation, as the following quote from J. S. Semancik illustrates: "Some form of stem slashing with razor blades or pricking with needles has proven most effective as a general inoculation procedure."

Such inoculation occurs not just in the laboratory. For example, infectious RNAs have been transmitted from one plant to another via contaminated pruning devices, leading to wide "epidemic" destruction, although simply breeding a contaminated stock has sometimes done the same. This certainly is reminiscent of the mechanical entry of a human retrovirus into a person (from intravenous drug use, blood transfusions, and other medical interventions) and of the transmission of the bovine leukemia retrovirus by veterinarians performing a series of vaccinations with the same contaminated needle. Thus, just as we can view some of the spreading of new kinds of virus diseases in animals and humans as problems of the twentieth century, we can also consider in the same way the movement of viroids and their devastation of plants. (One disease called codang-codang is said to have killed about a half-million coconut palm trees in a year.)

The mechanism of viroid-induced disease in plants is not fully known, but some insights into its method of reproduction are becoming clear—for instance, it uses an enzyme produced by the plant cells to replicate RNA (RNA → RNA). Because human cells do not carry this kind of enzyme, we should be spared an invasion by viroids, at least if this requirement is absolute.

But recently a structure containing a viroid-like RNA has been identified in human cells. For several years we have known of a defective virus called the Delta agent, which needs hepatitis B virus as a helper to replicate itself, and which by itself can cause hepatitis. Scientists at Rockefeller University and at Georgetown University have recently shown that the RNA of this defective virus has a form extremely similar to the peculiar structure of the viroid of plants. Although the Delta agent has protein and a virus-like structure (and therefore cannot really be considered a viroid), the fact that its RNA is related to that of a viroid suggests that we have not reached the end of variety of disease-causing structures.

Whether Delta agents can be made into transmissible pathogens in the naked RNA form by some inadvertent human intervention that allows them to enter our bloodstream is an intriguing thought. I have noted that human cells do not contain the enzyme that catalyzes RNA → RNA, so we should not have to face this problem. Yet, it is not inconceivable that a second microbe, carrying such an enzyme, could act as a helper, directly replicating this RNA as a viroid, or that some human misadventure might lead to further loss of control of a class of pathogens I will discuss shortly, called *prions,* which are proteins without nucleic acids! If either occurs in the near future, medical science will be hard-pressed to provide an effective response.

In an effort to stimulate the search for viroids in animal cells, some leading scientists in this field suggested that such agents might be the best candidates for the cause of scrapie in sheep and of kuru and related brain diseases in humans. These diseases were known to be transmissible by something smaller than a virus, something that, unlike viruses, is resistant to enormously high temperatures and normally destructive chemical and radiation treatments. Viroids appeared to be the prime suspect. But the actual agent was not a viroid but a prion, which survives extraordinarily hostile conditions (for example, temperatures of 280° C) because it has no genetic content—no nucleic acids, neither DNA nor RNA—and because of special features of the protein.*

Current evidence indicates that the introduction of these special proteins into the wrong cells at the wrong time is in fact what triggers development of the severe neurological disease called scrapie in sheep and kuru in humans. No one yet knows the exact mechanism by which prions replicate, but several interesting theories have been proposed. For example, since prions are proteins produced from a normal but

*Important contributions on prions have been made by several groups. I am aware mostly of those made by Stanley Prusiner and his colleagues at the University of California, San Francisco Medical Center, and by Carleton Gajdusek and his co-workers at the National Institutes of Health.

modified cell gene, when they are introduced into some cells, they de-repress genes for themselves (turn off repressor genes) and in so doing promote their own production. Sporadic cases of both naturally occurring prion disease and that associated with human intervention in tissue of the central nervous system are known.

According to Carleton Gajdusek, the latter include rare cases associated with corneal transplants, deep dental root-canal work, implantation of cerebral electrodes, inoculation of human growth hormone (made from human brain extracts), and brain cannibalism. We have learned about iatrogenically induced (medically caused) prion diseases from the earliest studies of Gajdusek and his co-workers on kuru, for which he received the Nobel Prize even before the strange nature of the agent was fully appreciated. The structure of the prion protein has been elucidated. It can interact with other molecules in a manner that forms structures similar to those seen in Alzheimer's disease and in other senile brain disorders, though in these disorders it is not believed to be from abnormal entry of the protein as in a "virus" infection but instead from an inherent abnormality in a cell protein that develops over time, perhaps genetically or environmentally stimulated to do so.

In recent years an epidemic has taken off of an apparent prion disease of cattle, substantially affecting the British cattle industry but now extending elsewhere. Apparently, similar mechanisms and similar pathology to kuru and related diseases of humans have been found. Not surprisingly, this epidemic seems to be a product of modern times. Gajdusek and others believe it is due to bovine and sheep bone meal fed to the cattle—a form of cattle cannibalism.

Even more diverse than the number of pathogenic microbes are the many mechanisms by which these agents, especially viruses, cause disease. Many simply destroy the cell or bring on its steady degeneration. Different hepatitis viruses, for example, destroy liver parenchyma; rhinoviruses bring on the degeneration of nasal epithelial cells; Venezuelan encephalitis viruses burst nerve cells; rotaviruses bring on the degeneration of intestinal mucosal cells; smallpox (vaccinia) viruses cause the death of epidermal cells. The signs and symptoms of infection with them correlate with these effects.

Viruses have also induced cell growth, exactly the opposite of cell destruction. In laboratory experiments several viruses have been shown to transform normal cells into those with properties of a cancer cell. For instance, the class of papovaviruses includes viruses that cause morphological (shape) alteration in cells and other effects that mimic tumor cells. Infection of some human blood cells (some T lymphocytes) with the retrovirus HTLV-1 leads to the permanent growth of these cells in laboratory tissue culture (a case of the previously described immortalization). Infection of other cell types (some B lymphocytes) with the herpes virus EBV (Epstein-Barr virus) also leads to the immortalization of these cells.

The indirect effects of microbes are even more diverse. For instance, the entry of Epstein-Barr virus into the B lymphocytes can trigger the abnormal growth of certain T lymphocytes *as a response* to the presence of Epstein-Barr proteins in B-cells and epithelial cells. This very response, overproduction of T lymphocytes, results in a common but usually self-limiting disease called infectious mononucleosis.

Consider also the diphtheria bacterium. Children infected with it often exhibited a bacterium-laden membranous material in the throat, but it is a toxin liberated from the microbe that caused brain damage. Many bacteria, and some other microbes as well, can cause pneumonia, but the chief pathological changes may be due not to the direct effect of the microbe on the infected cell but rather to the infiltrated but uninfected cells responding to the microbe.

The pathogenic effects of microbes can be even more indirect. Sometimes antibodies made in response to an infection will connect to some component of the microbe and produce antigen-antibody complex. When such complexes are present in large amounts, they sometimes produce skin abnormalities, kidney disease, blood disorders, and other medical problems.

Another example of an indirect mechanism is the well-named phenomenon of *autoimmunity.* There are many ways in which a microbe can induce some component of the immune system to react against the host rather than against the microbe. Many unexplained diseases of humans, including several forms of arthritis, diabetes, and some diseases of the brain, are thought by some medical scientists to be effects of such "leakiness" in the otherwise sophisticated armamentarium of our immune system—an attack by our immune systems against ourselves.

Finally, we do not know every angle of every mechanism for the majority of pathogenic effects of microbes. This fact will be pertinent to my consideration in chapter 15 of Koch's postulates and arguments raised against human immunodeficiency virus (HIV) as the cause of AIDS.

II

The Discovery of Cancer-Causing Retroviruses in Humans

4

The Story of Retroviruses and Cancer: From Poultry to People

Cancer-causing viruses in mammals were not found until decades after their discovery in chickens at the turn of this century. In fact, the field of retrovirology—the study of RNA tumor viruses—did not really develop until the 1960s. There are several reasons for this delay.

In a series of experiments on chickens, Peyton Rous found in 1910 that inoculating a healthy chicken with tumor tissue from a chicken with a sarcoma could rapidly induce a tumor in the healthy chicken (within a few months). But when other researchers tried to repeat the experiments, using mice, rats, and other mammals, they could not replicate Rous's results.

We now know that sometime during its history of passage from animal to animal, the retrovirus involved in the chicken sarcoma acquired a new gene from the DNA of a cell it had infected long ago. This special gene, called an *oncogene,* when incorporated into the genetic information of the virus, gives the virus a rare property: the ability quickly and completely to convert a normal cell into one with all the properties of a cancer cell.

Fortunately for animals and humans, such viruses originate rarely and are almost always defective. Most of these rapid cancer-causing viruses can neither infect host cells nor reproduce without a helper virus, which researchers must usually provide to maintain the oncogene-containing virus in the laboratory. Other than in such laboratory experiments, cancer is almost never transmitted by this kind of retrovirus. Because of the peculiar rapidity of induction of this chicken cancer (unlike what seemed to occur in human cancers), because clinicians saw no evidence of cancer as a communicable disease, and because Rous's findings could not be easily replicated, and had never been produced in mammals, the notion of viruses as cancer-causing agents in animals (or in humans) fell into disrepute.

Still, a few people persisted. By far the most important work was done by Ludwik Gross, in the early 1950s. This Jewish Polish refugee, working patiently with mice in small quarters at the Bronx Veterans Administration Hospital, was determined to find out why Rous's results with chickens could not be produced in other animals. For years he tested possible reasons: viruses might be more effective inducers of cancer in the newborn than in the adult animal; the time needed for tumor induction might be inordinately long; tumor viruses might be vertically transmitted—that is, by congenital, *in utero* infection or by newborn infection—making epidemiological data for communicability impossible to get. His first tentative successes of transmitting tumors suggested still another possibility. Some inbred strains of mice might be genetically more susceptible than others.

Gross's work was greatly aided by the breeding of such mice at the Jackson Laboratories in Bar Harbor, Maine, and by the contributions of a few other researchers, particularly those of another New Yorker, Jacob Furth of Columbia University, and John Bittner, whose work led to more evidence for the existence of mouse tumor viruses, including an agent that seemed to be involved in the development of breast cancer in some strains of mice.

But it was Gross who really opened the field, being the first to show conclusively that viruses could induce leukemias and lymphomas in mammals. The virus that induced tumors in Gross's mice, and the Bittner agent that induced breast cancer in mice, were of the same general type as the one that had induced sarcomas in Rous's chickens. A few years before, Rous, Vilhelm Ellerman, and Oluf Bang described, but did not characterize, a leukemia agent present in extracts of chicken leukemic cells. Later work demonstrated that these, too, were of the same type. Soon they would be referred to as RNA tumor viruses, or sometimes simply as leukemia viruses, as this was the most common form of cancer they produced. Later they would be called retroviruses.

It took roughly ten years for Gross's work to be fully appreciated. In fact, I am told that in its early phases it was frequently ridiculed. But in that interval other scientists joined the chase. Following Gross's lead, they examined different strains of mice and different types of leukemia and succeeded in discovering important variants of the Gross leukemia virus. Each new strain of the mouse virus could induce one or more tumors, but not every one resulted in leukemia. Named for the individual who isolated the retrovirus, some examples of these are Moloney leukemia virus, Rauscher leukemia virus, Friend leukemia virus, Kaplan radiation leukemia virus, and Rich leukemia virus. Almost at the same time, viruses more similar to the chicken Rous sarcoma virus than to a leukemia virus were also discovered in mammals: the mouse Moloney sarcoma virus, the Kirsten sarcoma virus, and the Harvey strain.*

The next major advance in retrovirology came in the early 1960s, not really out of "pure" basic research or human clinical research† but from the work of veterinarians. This research established that transmissibility of cancers by viruses was not limited to laboratory experiments. William Jarrett of Glasgow found a retrovirus of cats (feline leukemia virus) that could produce leukemias and some other kinds of cancer. Oddly enough, Jarrett also showed that the same virus could induce abnormalities of the immune and blood cell systems that resulted not in abnormal excess cell growth, which occurs in leukemia and other cancers caused by the same virus, but in the premature death or undergrowth of these cells—almost the opposite of a cancer! These findings were not lost on us, as we later considered this type of agent as a cause of the immune-deficiency disease later known as AIDS. But,

*This curious habit of naming almost every isolate of every retrovirus for its discoverer fortunately ended sometime around the discovery of the cat leukemia viruses in the 1960s.

†I refer to *basic* research in biomedical sciences as those areas of research that deal with the most fundamental questions in biology, usually without regard to any application or even to consideration of benefiting in any way. Generally, such research exploits the simplest possible biological systems, as in the early years of molecular biology's humands focus on the simple viruses that infect bacteria (the so-called bacteriophage). By contrast, *clinical* research involves people, or cells and tissues derived from people, and seeks to help people; it is *applied.* Ironically, it is not so rare that helpful medical advances come out of basic research. Sometimes I appreciate more the definitions given by the famous medical scientist and author Lewis Thomas. He says basic research is that in which the answer is never predictable, while applied research has predictable answers. As such, what I have called "basic" would often be "applied" and, paradoxically, clinical "applied" research often "basic."

in my mind, the major implication of Jarrett's results was that retroviruses cause cancer and certain other diseases in animals under natural conditions outside the laboratory.

This would soon be verified and extended by Myron (Max) Essex of Boston, William Hardy of New York, Oswald Jarrett of Glasgow (the brother of William), and Ed Hoover of Ohio, whose results supported the idea that in addition to the vertical parent-to-progeny transmission noted in Gross's mouse work, this kind of virus also moved horizontally, from adult cat to adult cat, and might even be sexually transmitted.

As is not rare in the history of retrovirology, this highly important work was not at once widely acknowledged. It raised what many considered the absurd notion that some naturally occurring cancers, at least in animals, could be caused by infection—a notion that was not simply ignored, but scorned. Everything medical people and scientists had learned about cancers in humans suggested that they were not communicable from person to person. Also, this phenomenon was not seen in the animal retrovirus laboratory models most virologists studied, such as the mouse system. Yet the accumulating, but not widely known, work with the chicken leukemia virus (the virus first detected by Ellerman and Bang) also suggested that apparently under natural conditions this virus could cause leukemia in chickens by infection. Though this was usually by vertical infection (from hen to fertilized egg), occasionally it was by horizontal infection, as with any conventional virus.

In the late 1960s and early 1970s, the virologists Robert Huebner and George Todaro, then both at the National Cancer Institute, proposed a theory that seemed to belie the significance of the work being done by Jarrett and others in the cat system and of that on chicken leukemias. Huebner and Todaro proposed what came to be called the virogene-oncogene hypothesis to explain the origin of all cancer, and in the process a new term came into the vocabulary of retrovirology: the *endogenous* retrovirus.

Scientists now began calling most of the animal retroviruses found by Gross and by the veterinarians *exogenous,* as they came into cells from outside, as an infectious agent. They had to find a way to infect the cell, to reproduce themselves, and to cause disease. But then, studies of the chromosomal DNA of various cells of all tested vertebrates began to show something very interesting.

Sitting quietly along the DNA of cells from every animal studied—from chickens to mice to humans—seemed to be evidence of the complete or partial genome of at least one kind of species-specific retrovirus. In this DNA form, such viral sequences were named *proviruses* by Howard Temin. The origin of these proviruses and the reason for their survival as "normal" genetic elements in animal cells, passed on by the germ cells of the organism (sperm and ova) from generation to generation, just like any gene—for example, that for hair color—were simply unknown. They may have been there since the cell's origin—normal genes serving some useful purposes but slowly evolving into the precursors of today's retroviruses. Or perhaps these ubiquitous retroviral pieces were remnants of ancient infections of early life forms. As such, they would eventually become like normal genes. Only rarely were they expressed and only rarely, or under special laboratory conditions, could they be found to produce complete, freestanding RNA viral particles. Most of the time they seem to be inactive.

The trouble was that no one knew (or yet knows) whether endogenous proviral DNA elements now contribute some positive function to the biology of the host, whether as such they are ever involved in diseases of animals or humans, or whether they are evolutionary remnants of "junk" DNA. Huebner and Todaro suggested that all cancer was due to the turning on of this endogenous proviral DNA.

When Huebner announced his virogene-oncogene hypothesis, he had already worked for many years and with much success in the Infectious Disease Institute of the National Institutes of Health. His decision to move at this time to the National Cancer Institute, also at NIH, coincided with President Nixon's declaration of his "great war on cancer" in 1971.

Several years earlier, spurred in large part by work on exogenous animal viruses (both retroviruses—or RNA tumor viruses, as they were then called—and DNA tumor viruses*), the National Cancer Institute had formed its Virus Cancer Program. It would become a special, highly visible, well-funded program; as such, it would be subject to much scrutiny. Its first head, Frank (Dick) Rauscher, would soon become the director of the National Cancer Institute, to be

*Around the time of the finding of the RNA tumor viruses in mice, other virus forms (DNA tumor viruses) were discovered that could produce tumors in animals. Later, as we will see, some DNA tumor viruses were also linked to the cause of some human cancers.

succeeded as head of the Viral Cancer Program by his associate John Moloney (known for his isolation of and work with mouse Moloney leukemia virus and another mouse retrovirus, the Moloney sarcoma virus). Once Huebner and Todaro announced their virogene-oncogene hypothesis, the Virus Cancer Program pushed for funding for work on endogenous viruses while still supporting some continuing research on exogenous viruses.

For their part, and with some justification, Huebner and Todaro remained unimpressed by exogenous retroviruses as possible causal agents; by comparison to endogenous viruses, they were relatively rare. They continued to present their case for the ubiquitous endogenous retroviruses as the key villain. It was their activation from a quiescent DNA state to one in which expression of some sequences occurred, they argued, that would encode special cancer-causing proteins. Further, they suggested that this could occur with or without the formation of a complete virus. This last point was important because no one had yet proved the presence in humans of any freestanding retrovirus particle.

These endogenous retroviruses were, according to Huebner and Todaro, activated by chemical carcinogens, radiation, or even other viruses (including, in animals, those rare cases where cancers had followed infection by bona fide exogenous viruses). Within the genetic makeup of this endogenous retroviral material was, they suggested, *a* special gene, *the* oncogene, that was the parent of the cancer-causing protein(s) and might, indeed, be the key to all cancer.

Their theory was attractive, and it brought many scientists, particularly molecular biologists, into this field; now they could study the molecular biology of gene expression, a timely and appealing branch of molecular biology, while simultaneously entering the more visible and, as far as funding for research is concerned, also more lucrative career of cancer research.

Through much of the early part of this exciting period, I was still focused on my own work on transfer RNAs, which do their work outside the nucleus. It was largely because of my friendship with Bob Ting, then a well-established NCI investigator, that I first began even to think about viruses. He and I had known each other socially before becoming colleagues. We had met in 1966, shortly after I arrived at NIH. In the fall of 1967, while I was still working in Sid Pestka's group on tRNAs, Bob gave a seminar on the properties of

DNA and RNA oncogenic (cancer-causing) viruses—of animals, of course.

Bob comes from a large and prosperous Hong Kong family; his father was at one time physician to Chiang Kai-shek. But Bob had studied in the United States, having had the good fortune of doing his graduate and postgraduate work under two master scientists, both Nobel Prize winners: Salvatore Luria at MIT, who taught Bob about *phages,* viruses that infect bacteria; and Renato Dulbecco, who was at Cal Tech while Bob was doing his postdoctoral training there and with whom he studied animal DNA tumor viruses.

Dulbecco had made important contributions by working out ways to study the capability of some DNA tumor viruses to convert normal animal cells into cancerlike cells in laboratory tissue culture systems, and he made these tests quantitative, as had Temin and Harry Rubin with the Rous sarcoma virus. Before that, such studies, including those of Ellerman and Bang, Rous, Gross, and Jarrett, had been done chiefly with live animals.

Dulbecco's work allowed scientists to study the basic phenomena of a cell's gradual transformation from normal to neoplastic.* Even if viruses were unable to cause any cancer in humans—and at that time such was still the prevailing belief—these studies would be invaluable to our understanding of the cancer process.

Bob talked about many fascinating things during his seminar but, as he reminded me years later, I expressed particular interest in his observation that oncogenic RNA tumor viruses—that is, viruses such as the Rous sarcoma virus and the Moloney sarcoma virus, with these oncogenes incorporated into their RNA†—seemed able to transform cells *in vitro* within a few days and to induce tumors in chickens or mice within a few weeks. I was also interested in the fact that the same general class of viruses, the RNA tumor viruses, could cause leukemia in several different animal species.

In the weeks and months that followed, Bob taught me how to play tennis and introduced me to more advanced studies of Chinese food. He also listened endlessly as I talked about my experiments in human leukemias. He suggested that I use the virus systems he had learned

**Neoplastic* refers to new growth that is not normal, or simply excess, growth (as in a benign tumor) but cancerous.

†At that time the presence of an oncogene in these viruses was not known. In fact, no one knew why they transformed cells *in vitro* and produced cancers in animals so quickly.

from Dulbecco, explaining how one could grow normal, uninfected animal cells in tissue culture and convert a portion of them into tumor cells by the simple addition of a transforming, or oncogenic, virus.

If you wanted to look for a cellular change that was central to the cancer process, the argument went, it would be easier and cleaner to do it in this kind of controlled system rather than by trying to compare human cancer cells (such as leukemic cells) to their normal counterpart (normal human blood or bone marrow cells); the compared cells might have other, unrelated differences that would make it impossible to pinpoint the change.

I was not immediately persuaded, but in time I decided to continue my comparisons of normal human blood cells and leukemic blood cells while beginning to work with Dulbecco's animal viral systems. Thus began a series of collaborative studies with Bob Ting, using oncogenic DNA viruses, in which we looked for biochemical changes that occurred during the cancer process. Like many other investigators, we found several, but with each change the same question nagged: How could we distinguish meaningful changes from side effects having no real role in producing the cancer-related changes in the cell?*

While we were working through these problems, still using animal DNA tumor viruses, Bob began to tell me more about an entirely different class of animal viruses—the RNA tumor viruses. He introduced me to Huebner, both the man and his ideas.

For me, Huebner became an enormous presence, a dominating force, and a man of endless ideas. He was a large man of gentle humor, with cocker spaniel eyes. Having studied many viruses and isolated several new ones, he was in my mind one of the greatest of all the virologists. He had made his presence felt throughout virology, and by the late 1960s I understood that he saw his future as being in cancer research.

In science, one invested idea, though not correct, often brings dividends elsewhere. Huebner's major thesis was shown not to be true. Many things activate endogenous proviruses in animals without actually provoking a cancer, and cancers often occur without their activation. As for oncogenes, they do exist but not in the domain of

*It has since become apparent that transfer RNA changes are not central to the control of protein synthesis. It is very unlikely, then, that differences found in their amount and type have anything to do with conversion to a cancer cell.

the endogenous viruses; they are found at other chromosomal locations. In other words, endogenous viruses do not normally avail themselves of the presence of oncogenes, though in rare cases these genes are indeed captured by an exogenous retrovirus. This is, in fact, what happened in the case of the Rous sarcoma virus in chickens and of other viruses in mice, such as the Moloney and Kirsten sarcoma viruses; and it is the explanation for the extreme rapidity of cancer induction by these viruses. But more often the significance of such oncogenes in cancer lies in their abnormal activation by mutations or other genetic changes induced by carcinogens or developed by chance.

Still, in bringing many of us to think seriously about endogenous viruses, Huebner contributed, at least partly, to our thinking seriously about oncogenes and to the finding of several new oncogenes (there are now over thirty)—surely a significant contribution. The downside of the appeal of Huebner and Todaro's theory is that it led directly to a further lessening of respect for systems that implicated as part of the cancer process the infectious exogenous retroviruses—or, for that matter, any kind of infectious virus, including DNA tumor viruses—and too much effort was then expended on searching for endogenous retroviruses in animals and humans. Thus, lying in wait for an idea or data, or preferably both, were the questions: How do the infectious exogenous retroviruses, those not containing an oncogene, cause cancer? And were there more of them? Soon we would learn some answers to these questions. In the interim, we could note that such retroviruses did their harm long after the time of infection—a feature much more consistent with what we knew of the development of human cancers after exposure to chemical carcinogens and radiation.

But Huebner's work would have a dominantly positive result for me (as it would for others), raising my curiosity about viruses in general and causing me to read broadly at a time when veterinarians were making enormously important contributions in exogenous infectious retroviruses in animals. It is not often that the work of veterinarians provides so rich a vein for investigators into human disease, but in this case it did. There in the world of some of the foremost veterinarian scientists I discovered the seeds of the idea that humans might also be pathogenically infected by these kinds of viruses.

The work that influenced me came from William and Oswald Jarrett of Scotland and later Essex, Bill Hardy, Ed Hoover, and others (each of whom did work on the cat leukemia virus); Peter Biggs in England and Creighton in the United States (chicken leukemia virus); and Janice Miller, Carl Olson (United States), and V. Van der Maaten (United States) and later Arsene Burny of Brussels (cow leukemia virus) and Tom Kawakami (gibbon ape leukemia virus), as well as from some of the earlier writings of Ludwik Gross and others regarding mouse leukemia. Of course, I needed no reminder that despite all the past claims by electron microscopists and other virologists positing the existence of human tumor viruses, it was for virologists a blind trail, a graveyard of experiments. Still, I wondered about the possibilities, though I did not immediately work on them.

For the most part, I was now comfortable doing my comparative biochemistry between normal and transformed cells. There were also the experiments on tRNA but their study was beginning to decline in my group. By 1968 or 1969 I began to study enzymes that catalyzed the making of DNA, called *DNA polymerases.* These are part of the process of cell division and are also needed to repair damaged DNA. It was a reasonable possibility that one of these enzymes might prove important to the abnormal growth of leukemic cells. I studied them and again I compared.

But somehow I was not satisfied. It seemed increasingly probable that nothing really important would come out of all this comparative biochemistry, at least not from me; and I was not alone in my frustration. Most investigators in cancer research at that time puzzled over which experiments had any real meaning. The door of the house of molecular biology may have been opened, but most of us could not see in to decide which room to enter.

In early 1970, I ran across the great molecular biologist from Columbia University, Sol Spiegelman, to whom I conveyed my reservations about the work I was doing in comparative biochemistry. He also expressed the fear that tRNAs might not be critical to the control of protein synthesis. The action, he said, would more likely be in the nucleus, at the level of activation of different genes and the subsequent formation of different messenger RNA molecules. He strongly suggested that I drop this tRNA work. I did. But I could not decide what new work to pursue—cellular DNA polymerase or some new angle in molecular biology.

Houston . . . May 1970 . . . A warm day at a cancer congress and everything is changed.

I had, of course, heard about Howard Temin from Ting. He too had been a student of Dulbecco through the same period as had Ting, and had then moved on to the University of Wisconsin. From Ting I knew he was a very bright biologist. I also knew that the most famous part of his work was controversial. Later I would come to see him as one of the most insightful and important biologists of our time. During the 1960s Temin and a few others thought that RNA tumor viruses did something unique in all biology: they reproduced themselves by going through a DNA form. In other words, these viruses could somehow convert their RNA genetic information into DNA. This notion was met with almost uniform incredulity.

Chiefly because genetic information was known to go only from DNA to DNA or from DNA to RNA, some critics went further and ridiculed the experiments and the idea (called the *DNA provirus hypothesis,* because the DNA form was known as the provirus). But at the Houston meeting, Temin announced his discovery, with his associate Satoshi Mizutani, of a special DNA polymerase contained within the virus, which could make DNA from the RNA of the RNA virus! Temin called it an RNA-dependent DNA polymerase, but John Tooze, a clever biochemist and then writer of editorials, working with the journal *Nature,* provided the label that would become the basis for the name of this new class of virus. He called the enzyme *reverse transcriptase,* and the class of virus later became known as *retrovirus.*

This was electrifying news, made public shortly afterward by joint publications in *Nature* by Mizutani and Temin and, simultaneously but independently, by David Baltimore, a brilliant scientist who had been working on various enzymes in different classes of viruses. The disbelievers and scoffers were now reduced to a few diehards. Baltimore was then working at MIT and is now president of Rockefeller University. Both Temin and Baltimore would later win Nobel Prizes for the discovery of reverse transcriptase.

After some late-night discussions with Bob Ting, I decided to enter the fray. My own laboratory, then in its incipient stages and staffed by only a couple of technicians and a few postdocs, would immediately be set up to compare the properties of reverse transcriptase enzymes from many different animal retroviruses. The experience with DNA polymerases, cellular enzymes that catalyze the making of DNA in the cell, gave me an entry point in the field of retrovirology that would not necessarily have to be at the very bottom. After all, reverse transcriptase was an enzyme that catalyzed the making of DNA, therefore it too was a DNA polymerase, though one with a very special property. And I would continue to study the DNA polymerases of human cells closely; surely there would be characteristics common to both the cellular DNA polymerases and this reverse transcriptase. But I would also begin to do comparison studies of evolutionary patterns of reverse transcriptase in different retroviruses, and to look for evidence of cellular reverse transcriptase in a variety of animals.

Because viruses have co-evolved with cells, viral genetic (nucleic acid) components usually have some similarities to those of cells. It would be reasonable, then, to think that cells might contain a gene for an enzyme similar to reverse transcriptase, and that this enzyme might sometimes be found in cells. In fact, Temin had almost immediately made this suggestion, and even proposed new theories on the use of such an enzyme in cells. For example, sometimes a cell may need a gene in many copies instead of one copy. In fact, more than one copy of some cell genes had already been observed. Temin proposed that one function of a putative cellular reverse transcriptase might be using a messenger RNA molecule as a template and synthesizing its DNA copy, which would then be integrated into the chromosomal DNA. Repeating the process would give many such copies. In hitching my wagon as tightly as I could to this horse, not only would I characterize reverse transcriptase from many different animal retroviruses but I would try to find the cellular counterpart of the viral enzyme and, if successful, also try to characterize it from many different animal cells.

My second objective would take longer and would spring from a more fragile foundation. The older, accepted method for the first indication of the presence of a retrovirus in a cell was to take pictures of the cell through an electron microscope. Often these pictures did not confirm the scientist's hunch, even when the scientist was right, because many viruses reproduce in bursts and it would take extraordinary luck to be on hand with an electron microscope at just the moment of new production. Moreover, when tissue is put into culture dishes, often the virus is formed for only a short time. And the technique was not only insensitive—that is, it could easily miss finding a small number of virus particles—but it frequently led to mistakes because some normal cell structures could be confused with viruses unless considerable care was taken in distinguishing them. Also, electron microscopy had practical disadvantages for me, as we did not have an electron microscope in the lab. We had to work out a collaborative arrangement with other groups whenever we wanted to use one, which often turned out to be expensive, time-consuming, and inefficient. But the discovery of reverse transcriptase suggested that there might be a way around the problems of electron microscopy. If one wanted to look for evidence of a retrovirus, why not begin by first testing solely for the presence of this virus-specific enzyme—a "footprint" of the virus, if you will? This would be a simple test that could be performed repeatedly, easily, and inexpensively by what is called an enzyme chemical assay.* Although, in the beginning at least, we would frequently mistake cellular DNA polymerases for viral reverse transcriptase, within time we could probably make the assay more sensitive and more specific. As I have indicated, viral components often have their counterparts in cells. However, they are usually quite distinguishable from each other.

To the extent that the test for the viral form of the enzyme could be made both sensitive and specific, I would have, in effect, at least a preliminary test for the presence of a human retrovirus. I might still want to confirm my results with an electron microscope, but I would have a basic test, easily performed in my own lab, to begin looking for evidence of human retroviruses. This became my long-term objective.

The story of the next ten years is more than that of our lab's discovery of a disease-causing retrovirus in humans. It is also the story of how our lab—over many years and through many experiments, with the assistance of clinical collaborators and epidemiologists and aided by interlaboratory exchanges of information, technical know-how, and newly developing laboratory expertise—learned to marry white-blood-cell biology to the new developments in retrovirology. Before we could demonstrate the presence of a human retrovirus in

*Enzyme assays employ techniques of biochemistry. Many proteins function as specific enzymes, that is, they make possible highly specific biochemical reactions. These biochemical reactions can be determined and quantified. The presence of a given specific reaction and the rate of that reaction are a measure of the presence (or absence) and the amount of a given enzyme, the essential catalyst making the reaction possible. Many times these are measured by providing one chemical substance (the substrate) which is "tagged," adding the cell or viral material under test for the presence of the specific enzyme and following a change in the "tag." Sometimes the tag is a chemical that will undergo a color change. Today we usually use a radioactive tag in our test substrate. The assay for reverse transcriptase is complex as enzyme assays go. Essentially, the substrate(s) are the nucleotide building blocks for making DNA, at least one of which is radioactive. In the presence of RNA (acting as the template or guide) and only if the enzyme is present, DNA will be made. It will be radioactive and can easily be separated and the amount measured, reflecting the presence and amount of reverse transcriptase.

human leukemic white blood cells, we first had to figure out a way to keep a retrovirus growing in the laboratory in primary cells—that is, in cells taken directly from a patient. If this could be done, by learning how to grow different kinds of human blood cells, we would have a way to detect the virus using the reverse transcriptase assay. In this way, we could at least obtain some preliminary information on the characteristics of the virus. But because growth of blood cells might be difficult to do for long periods, later we would want to learn how to transfer the virus into other target cells, perhaps by co-culturing the virus containing primary white blood cells with a target cell that was an immortalized cell line of animal or human origin. To first grow human white blood cells, we had to find growth factors, and this work took us into the fascinating area of lymphokines (of which cell growth factors are a subset), biologically active molecules acting at a distance on unrelated molecules (though the name *lymphokine* was not yet in existence when we started this work). During the course of these studies, we also began studying human oncogenes and their possible role in the mechanisms involved in the formation of leukemias and lymphomas. Basic animal studies pioneered by others* suggested that no matter the cancer—whether from an infecting virus, a chemical or other environmental factor, or simply a chance genetic change—ultimately these genes might prove pivotal.

One such study from our group,† for example, led to the important discovery (also discovered independently by Phil Leder, then also at NIH and now at Harvard Medical School) that a major change in one such cellular gene called *myc* was likely to be the essential change leading to the development of a specific kind of B-cell lymphoma known as Burkitt's lymphoma, usually occurring in children in certain parts of equatorial Africa.

This tumor illustrates remarkably the multiple stages in the formation of some human cancers because its development would be exceedingly rare except for the interplay of the Epstein-Barr virus with a severe form of endemic malaria. They combine forces to increase dramatically the probability of the key change in the *myc* gene. In fact,

*A list of all the important contributors to the study of oncogenes cannot be provided here, but the interested reader can find many reviews on this topic. See, for instance, Robert Weinberg in *Scientific American*.

†Carried out chiefly by Riccardo Dalla Favera, then a first-year postdoctoral Fellow with Wong-Staal and myself and now a professor at Columbia University School of Medicine.

this was one of the first identifications of a gene important to human cancer. During this period we would also perfect many of those important laboratory technologies in white-blood-cell research that would prove so useful to us in AIDS research.

As the first year passed, further advances in animal retroviruses strengthened my commitment to search for a cancer-causing human retrovirus. The first of these came from the work of Carl Olson and Janice Miller and their co-workers at the University of Wisconsin, who discovered a leukemia-causing retrovirus in cows. It too would later be shown to be an infectious exogenous retrovirus, creating new interest in William Jarrett's work as well as in similar work by others on chickens and mice. Of great interest to me was the observation that the bovine virus was present in very small amounts, and therefore difficult to detect and to isolate. It also suggested that a widely held belief might no longer be valid: extensive virus replication, leading to what I have earlier defined as a viremia, might not be needed to induce leukemia—at least some leukemias. Possibly most important, it once again opened the door to the nagging possibility that one or more human retroviruses had been missed.

As might be expected, Olson's and Miller's findings were not readily accepted. There were conjectures that they had not made a real observation; because the virus was so difficult to detect, critics claimed they might be wrong. And there was another, more practical problem: How many respectable laboratory scientists would be interested in bringing cows into their labs to do confirmatory experiments?

Further fortifying my resolve to continue looking for human retroviruses were the isolations of retroviruses from different primates; there were several such isolations by the early 1970s. Most important were the isolations of infectious exogenous retroviruses of gibbon apes, which produced several forms of leukemia in these animals. This virus story proved particularly interesting to my colleagues and me because the viruses had been found in a species close to humans, and the leukemias they produced were very similar to human leukemias. This work also produced the first persuasive evidence for interspecies transmission of retroviruses. Molecular analyses of the gibbon ape viruses—by George Todaro, Raoul Benveniste, and their co-workers at NIH—showed that they were closely related to Asian mouse retroviruses! These and other results argued that the gibbon ape leukemia retroviruses were the result of some long-ago interspecies transmission from mouse to gibbon apes.

More astonishing were the findings with a virus isolated from two pets in the same household. One of the pets, a woolly monkey, had developed a cancer. From this cancer, Gordon Theilen in Oregon and his co-workers isolated what they believed to be a new retrovirus. Soon Fritz Deinhardt, then at the University of Illinois and now in Munich, showed the virus could transform cells in culture and produce sarcomas in some New World monkeys. The virus was named the woolly monkey virus, also known as simian sarcoma virus.* When my co-workers and I analyzed this virus, it turned out to be another gibbon ape leukemia virus, further suggesting interspecies transmission.

Although it was clear that the woolly monkey and gibbon ape viruses were alike, they were not precisely alike. In the genome of some of the virus particles taken from the woolly monkey, we found that a fraction of the virus particles contained a gene that matched one from the woolly monkey—thus we had found an oncogene—and we could provide evidence that it came from its new host, the woolly monkey!

Many groups, including ours, contributed to further unravel the relationship between cancers and viruses. In regard to the specific example of the woolly monkey, we learned that what had appeared to be a leukemia-causing virus in this genome produced not a leukemia but a sarcoma—a much faster-growing form of cancer, the same form that Rous was able to transmit so readily from chicken to chicken. The woolly monkey never lived long enough to develop a leukemia. The induction of leukemia by the non-oncogene–containing retrovirus not only takes much longer than the induction of sarcoma caused by the oncogene-containing retroviruses, the precentage of infected animals developing the cancer with the former is much lower, too.

The explanation for how a virus associated with leukemia became a virus causing a sarcoma would be given chiefly by Stuart Aaronson and his group, also of the National Cancer Institute, and Russell Doolittle and his colleagues at the University of California at San Diego. They showed that this new cellular DNA that had attached

*A *sarcoma* is a type of cancer, usually of muscle, bone, or connective tissue. This woolly monkey virus produced a sarcoma of connective tissue. Usually an oncogene containing retroviruses produces sarcomas, as with the tumor produced by the chicken Rous sarcoma retrovirus.

itself to the woolly monkey virus consisted of a gene for an already well known cellular growth factor: a protein called platelet-derived growth factor (PDGF) that promotes growth of several kinds of cells. PDGF has since become one of the well-known oncogenes and was named *sis* (from *si*mian *s*arcoma) by Wong-Staal and myself.

Needless to say, the second pet in this household, the source of the infection of the woolly monkey, was a gibbon ape. The scientific community was now more accustomed to the notion of infectious retroviruses of primates. As well, we would now be prepared to think possible an interspecies transmission of an animal retrovirus into humans. Gibbon apes are, after all, relatively close to humans on the evolutionary tree.

As these discoveries were coming to our attention, we and other labs,* each working independently, continued the slow and careful process of becoming familiar with the reverse transcriptase (RT) enzyme from all the known animal retroviruses—in order to be able to distinguish each from the DNA polymerase enzymes normally present in human cells (which we would continue to characterize) and, of course, to make the assay (or marker) for RT as sensitive as possible.

By the early 1970s, once we had determined the best assay conditions for a given RT, the assay would be much more sensitive in its discriminations than electron microscopy (EM), the older way of identifying any virus. With a newly discovered virus, however—and this would be just as true today if a new retrovirus suddenly appeared—the first job would be to play around with the assay conditions to find the best combination to search for the virus's RT. (*Assay conditions* refer to the concentration of various ingredients of the RT enzyme reaction mixture as well as the pH, buffer, salt components, and time and temperature of the procedure.) Because we had not yet isolated a human retrovirus, we were compelled during this period to try a wide range of assay conditions.

In 1972, it seemed that our patience had finally paid off. Due chiefly to the work of two of my co-workers, M. Sarngadharan (in the lab he is known as Sarang) and Marvin Reitz, we had reason to believe we

*In particular, these included Sol Spiegelman and co-workers at Columbia University; Maurice Green of St. Louis University; and the discoverers of reverse transcriptase, Howard Temin (with his then new student Yung Kang, now head of microbiology at Ottawa University) and David Baltimore.

had our first clear-cut evidence of a retrovirus in human cancer cells.

Sarang, who was born in Kerala, India, had joined the lab just a year earlier to work on RT assays. Reitz also came to us in 1971, as a young postdoc in biology. Together, they obtained a sample of cells from a person who had a leukemia of the lymphocytes (lymphocytes being a form of white blood cell). Not fully appreciative at this time,* we were probably looking at precisely the kinds of lymphocytes we would later learn are targets of human retroviruses. We were delighted but surprised to find that we had purified an enzyme with all the then-known properties of a viral RT.

With more than a little excitement, we were ready to publish our findings, submitting an article to the prestigious and widely read journal *Nature.* Word came back soon after that it had been accepted. We were elated. Here at last was vindication of our beliefs that a human retrovirus was indeed present in human leukemic cells. We eagerly awaited the response from the scientific community to the publication of our findings.

To the puzzlement and dismay of all of us, myself most of all, our findings aroused little interest. Once past our initial disappointment, we soon figured out what had happened. This had not been the first time results were published suggesting the presence of a reverse transcriptase in a human leukemic cell; in previous publications, other groups, including ours, had—with as much enthusiasm, although with far less clear-cut findings than we had now—claimed the same finding, only to have doubt cast upon it with further investigation.

It was easy in those days to confuse viral RT with one of the normal cellular DNA-synthesizing enzymes (what are called DNA polymerases), and most of the early viral RT findings were rejected as probable cellular DNA polymerase enzymes. In fact, in a case of ours two years earlier, the results were clearly noted as being of a preliminary nature, and our own interpretations included the possibility that this was a cellular DNA polymerase, which had some reverse transcriptase–like properties. But our 1972 findings were different. The RT was clearly not from any of the known cellular enzymes; it had all the characteristics of reverse transcriptase. Still, few seemed willing to listen.

Our first thought was to look for additional examples to support our contention. In the following years we did make a few similar findings.

*The technical capacity for defining lymphocyte subtypes was in its infancy.

But in time we had to accept the fact that although these few positive results fortified our own resolve to continue the search for these viruses in human cells, this approach was doomed to be seen as inconclusive. Finding a putative footprint (in the form of an RT enzyme) of a human leukemia virus would never be a substitute for isolating and growing the virus itself in a laboratory cell culture.

To do so would require major efforts in cell culture of the various kinds of human blood cells. We would need to grow various human blood cells routinely for long periods and probably in large amounts. But by this period (1973–74) only a small subset of B lymphocytes could be grown in long-term tissue culture, and those unpredictably. These successes were with the relatively rare B-cells infected with EBV, the herpes DNA virus discovered by Tony Epstein in the 1960s and identified as playing a role in Burkitt's lymphoma and as a direct cause of infectious mononucleosis.

Partial success had also been obtained with one other blood-cell type. Leo Sachs of Israel and, slightly later, Donald Metcalf and his co-workers in Australia described a method for the growth of myeloid cells based on their discovery of a protein (growth factor) that induced growth of these cells.* This new system for myeloid cell culture, however, allowed growth of these cells only in small numbers, for short periods of time, and in small "colonies" of cells on a solid surface. As important and pioneering as this work was and still is, these were not conditions that would be useful for our objectives. What we needed, ideally, was to find a growth factor that would allow us to keep the primary cells we were working with—leukemic white blood cells—continuously growing in solution.

The task of finding such a growth factor fell to Robert Gallagher and Zaki Salahuddin. Zaki had joined the lab in the early 1970s as a senior technician. Gallagher sagely suggested that our best shot at finding such a factor might be by working with human embryo tissue. Leukemic cells, as immature white blood cells, might express receptors for species-specific growth factors that normally were expressed only during fetal life. I agreed. Fortunately for us, restrictions on the use of human embryonic tissue for such purposes were not imposed until much later.

While Gallagher and Salahuddin searched for a growth factor, Sa-

*Growth factors for epidermal cells of the skin and for certain nerve cells had been discovered for the first time somewhat earlier by Rita Levi-Montalcini of Rome and Stanley Cohen of Vanderbilt University, for which they would receive the Nobel Prize in 1986.

rang, Reitz, and by then others, including Gallagher, continued to search for other leukemic cells evidencing viral reverse transcriptase.

White blood cells derive from one of two lineages: myeloid or lymphocyte. Those of the myeloid lineage include granulocytes (the pus cells that form in response to an infection) and the monocyte-macrophages, which play a role in inflammatory responses and in many other complex acts of immune defense. B-cells and T-cells are of the lymphocyte lineage. It was at about this time, in fact, that scientists began to be able to distinguish lymphocyte subtypes. The T-cells are further subdivided into types that can be distinguished by function and by the presence or absence of certain proteins on their surface. The T4 helper lymphocytes, sometimes called the pilots of the immune system, regulate the functioning of their own kind, as well as of other blood cells, mostly by secreting certain proteins known as lymphokines. Lymphokines produce interesting biological effects on other cells whose surface receptors admit them. Some lymphokines are growth factors. The parent, or stem, cell that gives rise to all these different lineages of white blood cells is present predominantly in the bone marrow, as are many of the earliest precursor cells of the different lineages of white blood cells, as well as the red-blood-cell precursors. The final, mature functional cells are widely distributed in the blood and easily accessible.

Leukemias are malignancies of white blood cells, and there are as many kinds of leukemias as there are kinds of white blood cells. In fact, there may be more. A leukemia can involve one cell type at one stage of its differentiation and a second at another stage. The stage of the involved cell determines clinical features of the leukemia, so leukemias that involve the same cell at different stages are each given their own names. Does the fact that the clinical picture differs mean that the disease, the cause, or the mechanism must be different? We did not know.

The difficulties were not made easier by the knowledge that leukemia of animals can be induced by several agents in the laboratory—for example, radiation, viruses, some chemicals, genetic mutations arising spontaneously, or certain genetic inbreeding. In other words, should we have been seeking *a* cause for a disease when we were not sure whether it was really one, three, or ten diseases, even when we had only one cell type involved and similar clinical pictures? As a further complication, experimental studies on animals leave no doubt that one type of leukemia can be produced in the laboratory by more than one agent.

Our requests to our clinical collaborators were for any and all kinds of tissue from patients with leukemias and lymphomas. Infrequently and sporadically, these cells tested positive for RT activity. But none produced evidence of RT activity as clear as that produced from the cells of the young man with adult lymphocytic leukemia, our first virus-like RT finding. I did not conclude that we ought to focus on this specific kind of cell leukemia, although perhaps I should have. I felt at the time that this positive result was due simply to better luck or, perhaps, technique. (In time we realized this patient had probably had the kind of leukemia from which we would later isolate the first human retrovirus.)

We continued to work with white blood cells from all forms of leukemia, those of both the myeloid and the lymphocyte lineages. Then, one day in 1973, Gallagher and Salahuddin succeeded in extracting a growth factor from the culture fluid in which one of their embryo tissues had been growing. With regular infusions of this factor, we could keep primary myeloid leukemic white blood cells in continuous growth. This was an exciting result. It was at this time that Steve Collins came to work with us as a new postdoctoral Fellow and joined in this work.

Then came the real surprise. For reasons no one understands, one specimen of those primary myeloid cells Steve was working on became immortalized—that is, became independent of any growth factor, the cells reproducing themselves without any help from the scientist except the usual cell culture nutrient fluid (called *media*). The first human myeloid cell line had been developed! This particular specimen never yielded any virus. Naming it HL (Human Leukemia)-60, we immediately made this myeloid cell line available to other scientists. It remains in use throughout the world as one of the most valuable of all cell lines for many kinds of biological studies of these kinds of cells.

Of even greater interest to us was another leukemic myeloid cell specimen, which we named HL-23. It followed the expected pattern in regard to growth, remaining strictly dependent on the continued presence of the growth factor. It never became an immortalized cell line, but these cells did test positive for RT activity. After careful monitoring, we found that they continued to be RT-positive. Now there was great excitement in our laboratory, particularly when electron microscopy confirmed a positive picture of the presence of a type-C retrovirus, the exact type of retrovirus that causes leukemia in many animals. We seemed to have exactly the kind of retrovirus we expected: one we immediately found was related to the leukemia-causing type-C retroviruses of primates, especially the gibbon ape virus. Yet, significantly, David Gillespie's early analysis of it seemed to indicate it was different from the gibbon ape virus.* We submitted our findings for publication.

But disaster loomed. Most scientists firmly believed at the time that human retroviruses did not exist. The burden of proving that they did rested squarely on our shoulders. Our job was further complicated by a tension in the scientific community that was not limited to the question of human retroviruses. Rather, it revolved around the clear proof of a relationship between viruses and cancer.

If human retroviruses existed, they would be capable of causing cancer—that's why we were looking for them in leukemia cells. And if some, any, cancers were caused by retroviruses, there might be many others that would then be found to have viral causes. For years stories had circulated in the lay press that scientists were about to discover that viruses caused cancer, meaning *all* cancers. The public thus expected the long-awaited cure or treatment for cancer. On a more sensible level, investigators had some time back presented evidence for a specific one-to-one relationship of a few viruses and a few human cancers—for example, the DNA herpes simplex virus and cervical cancer and for EBV (which, as we have seen, is also a herpes virus) in Burkitt's lymphoma.

But very soon the positive findings regarding herpes simplex were going the other way, leaving no clear evidence of its involvement in cervical cancer and, though EBV was (and still is) accepted as playing a role in Burkitt's lymphoma, it has become less important because it appeared to be only one of several factors needed for this tumor. Also, Burkitt's lymphoma is geographically limited to one area of Africa, so that its significance was downplayed. Finally, there had been many

*Gillespie had also joined the lab in the early 1970s as it was expanding. As a graduate student with Sol Spiegelman at the University of Illinois, he had co-authored a classic paper with Spiegelman, one of the seminal pieces of work on the development of the powerful technique called molecular hybridization. This technique allows a scientist to determine whether one nucleic acid is related to another, and even to fish out a particular gene from a sea of genes using a related piece of nucleic acid as a probe.

recent claims for human retroviruses, including some by very well known and respected virologists. In the end these turned out to be animal viruses arising as contaminants of cell lines.

It was at this time that we came along with our evidence of this new retrovirus as a possible cause of leukemia. The disaster that would soon wash over us, coming, as it did, on top of these many other negative findings, would end for many scientists further interest in pursuing the question of whether viruses play any role as causal agents in any cancers. Cancer virology came to be concentrated almost entirely on the rapidly progressing and important studies of oncogenes. Animal retroviruses continued to be studied, but only as an aid in understanding the biological mechanisms of cancer in animals, not as models for possible disease causation in humans by similar kinds of viruses.

My personal reaction to the disaster, which would place human retrovirology, and me with it, at a very low point, although obviously unpleasant for me to recall, is a necessary part of this story.

5

Success, Defeat, Success

It was after a restful and happy long holiday weekend at my parents' home in Connecticut that I first heard that the freezer storing Gallagher and Salahuddin's growth factor had been found unplugged. The magnitude of the problem was immediately evident: without the growth factor to feed the primary leukemic cells of HL-23, the cells would be likely to die or, at minimum, permanently stop growing. We would lose the virus.

We had taken the normal precautions against loss of the growth factor, having stored in frozen solution both an extensive stock of the original embryonic cells that produced the factor and a large number of individual vials of the already extracted factor. The fluid containing the growth factor could be thawed and used at any time. When it was depleted, or for any reason damaged, we could simply thaw the freezer stock of embryo cells, regrow them, and obtain the media in which the cells were nourished. This media would be rich in our valuable growth factor. Perhaps, in retrospect, we should have stored these stocks of frozen growth factor and frozen embryo cells in separate freezers in anticipation of such an event. But I didn't plan so carefully then. I do—now.

Without growth factor we could not keep the primary myeloid leukemia cells alive or the virus replicating, which meant that we could not perform the follow-up work needed to determine whether this virus had in fact caused the leukemia of the patient. As important, we could not give our HL-23 cells and virus specimen to other scientists for independent examination.

As word of what happened moved through the laboratory, we each expressed different feelings. In the first moments my own reaction was one of anger: Had someone done this deliberately? If so, why? Or, more probable, was it just a stupid accident, such as the nighttime cleaning crew pulling the plug from its socket in order to plug in a vacuum cleaner? Then gloom set in. The loss was of immense proportions.

Within a day or so the deepest gloom had passed and a few of us resolved not to be beaten by this event. In other situations in my past, often, perhaps even usually, positives had come out of negatives. Now I told myself that maybe somewhere, somehow, something good would come out of this.

Since the time we had obtained the supply of embryo tissue from which we extracted the growth factor, the question of fetal research—now entwined with the abortion issue—had become a sharp public controversy, and human embryo tissue harder and harder to get. Even within the lab, we were not all of one mind about the ethics of working with embryo tissue. For example, Phil Markham, who had joined the lab several years earlier, and I had long discussions about the ethics of using embryos, even though both of us acknowledged the scientific value of such research.

We were still able to get some embryo tissue, and for close to a year we tried and tried to find a new source of our original myeloid cell growth factor. Finally, we had to face the possibility of never recovering the myeloid cell growth factor. In order to continue research with the virus, it would now be mandatory for us to introduce the virus from the primary leukemic blood cells into some other cells, preferably into cells that we and others would have no problem continuously growing, such as an already immortalized cell line. We could hope that the virus replicated in this new environment. The first cell line we turned to was obviously HL-60. It was of human origin; it was of myeloid white-blood-cell lineage; it had been carefully checked for any virus and was found "clean"; and it was ours. But we could not get the virus to grow there. We tried other cell lines as well, those few that were then available of human, as well as many more of animal, origin: for instance, we tried cell lines of mouse, rat, bat, dog, and monkey origin. Each time we looked for evidence of reverse transcriptase, and each time our findings came up negative. We were discouraged. Finally, Robin Weiss and Natalie Teich from London joined the lab. Working with a variety of animal cell lines, they soon produced what we had been unable to do for many months—sample after sample positive for RT. Remarkably, Salahuddin and Gallagher also started to have success at about the same time.

Had I been a little bit wiser then, I would have been worried about some of these results—not about the investigators, because I have nothing but the highest regard for them, but about the high amounts of virus production and the sudden virus release more typical of a cell that has been long infected with an established, previously cultured retrovirus than of a new infection of a cell culture. Why, after months and months of failure by Gallagher and Salahuddin with the same animal cell lines, had there been these quick successes? Had Gallagher and Salahuddin originally been faulty in their procedures? Or were we now not really finding evidence of RT from a human leukemic cell? I didn't ask those questions then. Instead, I continued to believe in our positive results. We had done it once, and so I assumed we had done it again.

We eagerly prepared to talk about our findings at the upcoming annual meeting of the Virus Cancer Program in Hershey, Pennsylvania. In preparation for the meeting, we began our own follow-up analyses and sent samples to other scientists for independent examination and verification.

As the time of the meeting approached, our own follow-up studies started to suggest what we had earlier failed to consider: the possibility that the retrovirus we were finding was a contaminant, or contaminants, of animal retroviruses also being studied in our laboratory.

Contamination is a very common problem in any virus laboratory where more than one microbe is being studied, particularly common with retroviruses in very active labs. But because we were so sure that contamination of our original human myeloid cells had not taken place—they had been *primary* human leukemic cells taken directly from the blood, and the electron microscopy done on those cells supported our position—we clung to the assumption that we once again had the real thing.

Although our early analytical results had suggested a problem and, by the time of the meeting, I was beginning to have my own doubts,

I was not prepared for what happened at the meeting. Several researchers, one after another after another, came up to announce that examination of the cells we had sent out for independent confirmation had revealed not one, but two—and in the case of one cell line, three—different animal primate retrovirus contaminations.

What surprised me were not the findings—as I say, I was already developing my own doubts—but the vehemence with which they were delivered. More than one speaker used our misfortune to ridicule the very idea of a human retrovirus. It seemed as if a special effort were being made not simply to point out our error but to put the final nails in the coffin of the study of human retroviruses. The joke became: "Human tumor virus? Or human rumor virus?" Indeed, the whole concept that retroviruses were truly capable of causing serious diseases in humans came to be trashed in the backwash. Other candidate human tumor viruses (those that are DNA tumor viruses rather than retroviruses) were not faring much better.*

Even now I have difficulty thinking back to that day. I would be subjected to far more extensive, personal, and even vicious attacks years later when I entered AIDS research; I would even have to endure a long and trying inquiry within NIH as a result of one journalist's bizarre and obsessively defamatory article. But nothing compared with the feelings that passed over me as I sat that day in Hershey, Pennsylvania, hearing not just HL-23 but much of my life's work—the search for tumor-causing RNA viruses in humans—systematically and disdainfully dismissed. And despite the fact that we had made other useful observations related and unrelated to retrovirology, the observations seemed to leave no positive impact at that time.

What made it so painful? Was it that the irreparable loss from the freezer "accident" had only then become real to me? Was it just personal pride, a feeling that I had embarrassed myself, my family, my lab? To use a line from Abraham Lincoln, also borrowed by Adlai Stevenson for his concession speech in 1952, I was too old to cry, but it hurt too much to laugh.

Ludwik Gross came up to me after the last speaker had finished and

*Oddly enough, after the evidence that our isolate was a monkey virus, several groups in Europe and at least one in the United States (Werner Kirsten's group at the University of Chicago) reported isolations of the same type of monkey virus from human cells that were, I assume, also contaminants.

quietly put his arm around me. He told me that if I believed in myself and knew that my ideas and the fundamentals of our work were on the right track, I should not worry. I should continue doing what I was doing. He reminded me that his work had once been ridiculed, as had Temin's, and that science was a tough game; one should not expect much kindness from critics in a competitive field. If it turned out that viruses were not involved in the cause of any human cancer, even the compassion of a good friend could not make it true; and if it turned out that viruses were involved, then all the derision could not make it untrue.

Still, several weeks passed before I started to come to grips with what had happened at Hershey. The damage had been not just to myself, but to every one of us working in LTCB and to researchers elsewhere who had invested time and effort working with human retroviruses.

Had I been a lone, independent researcher rather than head of a lab, I might have spent the months after the Hershey meeting licking my wounds. But there was my staff to consider; they were very shaken by what had happened. Their careers had also been put on the line. Years later Zaki Salahuddin, whose freezer had been the unplugged one, talked about the incident with Mark Caldwell, a professor of English at Fordham University in New York and a writer who was doing a story about the lab. Salahuddin told him that the lab was in a state of shock as a result of Hershey: "I'd just come to the lab from Bangladesh. I remember wandering around the lab thinking, what was going on, until one day Bob took the whole staff to lunch and told everybody to ignore the outside criticism. Our responsibility was to get back to the real work of the lab, including continuing to look for human retroviruses."

For the next few months, as Marv Reitz remembers, "the lab as a whole was by and large depressed." We all had the same thoughts. The fact that the cultures of the virus contained not one but three contaminants—woolly monkey virus, gibbon ape leukemia virus, and baboon endogenous virus—was almost beyond belief. Either HL-23 had been a complete artifact or a real human retrovirus had been missed. "It was at this time, however," continues Marv Reitz, "that we all had the pleasure of making the acquaintance of Hugo, a gibbon ape with a T-cell leukemia." This is a great aspect of science. There is always something new.

Like most cell and molecular biologists, those in our lab rarely have the occasion to work with live animals. This has spared us one of the more painful aspects of scientific research—the need to confront the problem of animal rights and human research. I have often been asked where I stand on this difficult issue. During my period of AIDS research I was apparently quoted as saying that if I had to eliminate every chimpanzee in order to find the AIDS virus or a cure for AIDS, I would do so. Of course, my remark referred to an extreme hypothetical situation and, as such, doesn't really give an accurate picture of my feeling. If I felt it necessary to further valid research goals, and if there were no other way to go about doing this work, I might use live animals—in as humane a way as possible. Fortunately, I rarely have to confront this problem. But there is also no doubt that on the few occasions, such as the one I am about to relate, when we have worked with live animals, I have had considerable concerns.

Phil Markham recalls the story of Hugo: "I don't remember the details, but we became aware of a suspiciously sick gibbon ape named Hugo in a primate colony on Halls Island off the shore of the southeastern United States. Arrangements were made through contacts at Litton Bionetics, which had a small gibbon ape colony and could house the animal in special containment facilities, to ship the sick gibbon to us under quarantine conditions." Hugo was near death when he arrived and very shortly thereafter he died—but not before he had won our hearts.

The Japanese scientist Tom Kawakami had just shown that several strains of retroviruses could be found in gibbon apes with various kinds of leukemia. Now, as Reitz recalls, "Hugo could give us something real to study, as well as possibly providing us with a new variant of gibbon ape leukemia virus. We could also try out the techniques we had been developing for five years, such as reverse transcriptase assays and molecular hybridization assays on a real primate, with a real leukemia, and a real virus. In short, we would perform a molecular autopsy.

"Our anticipation was not misplaced," Reitz recalls. "When different investigators tested a variety of his tissues, the results went off the map. Hugo was like a bag of virus. Virus production in his heart, for example, accounted for 10 percent of all the messenger RNA in that organ. I remember a luncheon meeting at a Greek restaurant in downtown Bethesda to discuss data of the Hugo group, as those of us working on his tissue samples were now called. Dr. Tad Aoki, from

Japan but working at NCI, who was conducting the electron microscopic studies, stood up to present his findings. He held up electron micrographs and, pointing to them, began telling us things like 'Whole surface of ape tongue completely coated with virus particles.' Other people in the restaurant began to look over at us uneasily, perhaps wondering if we were dangerous.

"But we had a good, productive time with Hugo, who indeed proved to have a new variant of gibbon ape leukemia virus. It gave us all a needed shot of confidence that we could at least find virus when it was staring us in the face. The only downside was that it reinforced the notion, then prevalent among retrovirologists, that when virus was present, and involved in leukemia, it would be in a setting of overwhelming virus replication. This, of course, did not turn out to be the case with human retroviruses when they were finally discovered."

Robert Gallagher and other co-workers continued to test dozens of new human embryo samples to see if any released the needed growth factor for our human myeloid cells. The results continued to be negative. Then a newcomer to our group, Doris Morgan, who had an unusual degree—in hematology technology from M.D. Anderson Hospital and Tumor Institute in Houston—came to me for her postdoctoral work and joined this effort. Rather than have her duplicate ongoing efforts with embryos, I directed her to a different source.

In the 1960s Peter Nowell of the University of Pennsylvania had discovered that phytohemagglutinin (*phyto* meaning "plant," *heme* meaning "red blood cells"), or PHA as it is universally called by scientists, a plant extract known for its strange ability to make red blood cells agglutinate, could also make some human lymphocytes obtained from the blood of healthy people metabolically change from "resting cells" into cells that divide once or twice in culture.

A few years earlier, I had wondered whether these PHA-stimulated cells (later shown to be lymphocyte T-cells) released any growth factors. I soon found that they did and that one of them was the same growth factor discovered much earlier by Sachs and Metcalf and now named granulocyte macrophage-colony stimulating factor (GM-CSF). Sachs and the Australian group had found that this factor was released by cells from a few kinds of tissues. For example, Sachs suggested that the major source of GM-CSF might be the fully mature end-stage myeloid cell, the granulocyte. The granulocyte cannot reproduce itself. Its "genetic program" is fixed to help fight any infections, especially by bacteria. Normally, it then dies within a week or two. Granulocytes are constantly being replenished, however, from the more immature precursor myeloid cells of the bone marrow.

GM-CSF made by the granulocyte would be logical. It would "feed back" and promote growth and differentiation of these earlier-stage myeloid cells in the bone marrow to replace the now terminal granulocyte, a nice example, theoretically, of keeping a "steady state." They had no evidence, however, and expressed no belief that T-cells (which, as we have seen, are lymphocytes, a separate white-blood-cell lineage) would be a particularly rich source of this factor. In fact, at the time this would have appeared illogical. Why would an unrelated cell, the lymphocytic T-cell, be involved in regulation of myeloid cells? (Later, I wondered whether one of the reasons I was interested in this question was my experience with erythropoietin, which, oddly, was produced by the kidney cells but acted as a growth factor for red blood cells.)

Actually, of course, there is much precedent for regulation of one tissue by a substance made at a distance by another tissue—this is the whole business of the endocrine system. The thyroid gland in the neck and the adrenal glands sitting on top of the kidneys produce molecules that enter the bloodstream and act on many cells all over the body far from or close to their source of production. But these endocrine glands are obviously specialized tissues for this very purpose, making and secreting regulatory molecules we call hormones, and these molecules are generally quite small. What was new with erythropoietin was that the kidney, rather than a hormonal specialized gland, produced the molecule; the target tissue was not multiple cells and tissues but a very particular cell (a red-blood-cell precursor), and the molecule erythropoietin was large—a protein. The situation with the T-cells was even more extreme. They were simply single cells that made regulatory molecules for other cells, and the regulatory molecules they produced were also large (proteins).

In 1974, I published these results with a postdoctoral, Joan Prival, and my new co-worker, Alan Wu, showing that human T-cells made and released GM-CSF. At about this time, others were describing similar substances coming from T-cells with apparent biologically significant effects in immune function. Indeed, in retrospect these were the first discoveries of what, as a group, we now call lymphokines. In 1975, I wanted to see whether perhaps these same T-cells also released our lost myeloid growth factor from human embryos. Our factor differed from the Sachs-Metcalf one because GM-CSF promotes growth of myeloid cells as small colonies (or groups of cells growing in a small pile) only on a solid agar surface and only for short periods, rather than promoting long-term growth of these cells in a liquid broth cell-culture system, which would enable us to do the right experiments.

I asked Doris Morgan to perform a simple experiment: take human blood or bone marrow samples from healthy donors (normal bone marrow and blood are, of course, both rich in all lineages of white blood cells, and we knew from our earlier work that the factor promoting growth of human leukemic myeloid cells also promoted the growth of normal myeloid cells); put them into a nutrient broth; stimulate the T lymphocytes in the sample with PHA; wait differing periods of time; then sample the nutrient broth* and test its ability to induce long-term growth of human myeloid cells. A positive feature of the system included the fact that both our growth factor (whose source was T-cells) and our target myeloid cells would come out of this one sample of normal human blood or bone marrow.

Morgan patiently mothered her cell culture and eventually helped us come upon a major discovery. As long as a source of this conditioned media was added to the PHA-stimulated cells, some of these cells kept growing. Once the fluid rich in growth factor was produced, Morgan could either separate it from the stimulated cells and store it for later use, or reintroduce it back into the sample, stimulating further the growth of certain other cells.

There was a surprise in store for us. With an ordinary microscope, we examined our target cells only to find that what we thought were stimulated myeloid cells were instead lymphocytes. In other words, the T lymphocyte–derived growth factor was stimulating lymphocytes, not myeloid cells.

At first I reacted flatly to this news. For obvious reasons, I thought that upon further examination the stimulated cells would turn out to be the rare but already well known EBV (Epstein-Barr virus) immor-

*When used as a source of biologically active molecules released from cells, such nutrient broth, obtained after cells have been growing in it, is usually referred to as conditioned media, or CM.

talized B-cell. Growth of such cells had been previously achieved by several groups. By microscopic examination, B-cells and T-cells cannot be distinguished. Soon, however, new tests capable of distinguishing B lymphocytes from T lymphocytes revealed that what we had were not B-cells but T-cells; T-cells had never before been grown in long-term culture. We had found a growth factor from T-cells for T-cells, a cell for which there had never before been a growth factor.

During the next couple of months we characterized our growth factor and, with others, named it T-cell growth factor. We reported these discoveries in 1976 in the journal *Science* and in more detail in the *Journal of Immunology* in 1977, as a result of more painstaking work by another co-worker, Francis Ruscetti, who had joined LTCB as a postdoctoral in the mid-1970s. By 1980 I reported on its purification with another postdoc, James Mier.*

For my co-workers and myself, the fact that we had found a T-cell growth factor perhaps should have rung a bell. Our first clear viral-like RT positive assay—the basis for the 1972 *Nature* article by Sarang, Reitz, and myself—had been from the cells of a patient with leukemias of lymphocytes. But back then, we could not really further characterize those cells to determine whether they were B-cells or T-cells, partly because the technology to distinguish the two had not yet been fully perfected. Also it could not be done at the Houston hospital where this patient was being cared for. In hindsight, this patient fits well with the description of an adult T-cell leukemia.

In 1978 a new clinical postdoctoral Fellow joined us. Bernard Poiesz had spent one year in the medical wards at the National Cancer Institute. As a student he had gained some research experience with enzyme assays, and now he was eager to return to the laboratory bench. He was interested in our new T-cell growth factor. Mostly, he wanted to join our efforts in finding human retroviruses.

Poiesz's decision to join our group, and his enthusiasm for what we were doing, provided a needed lift to all of us. "From the

*But soon after its discovery in 1976, I brought this work to the attention of clinical immunologists, particularly Steven Rosenberg of NCI. T-cells are of great importance to immunologists because they play a vital role in the immune system, so this discovery received much attention. T-cell growth factor would soon be renamed Interleukin-2 and would find its way into most immunology and clinical oncology labs. There it would become an important tool in the therapy of many human cancers, an approach extensively developed and brilliantly exploited by Rosenberg at the National Cancer Institute. T-cells that attack cancer cells could now be grown in large amounts and given back to patients, a promising approach for cancer therapy.

early seventies on," Sarang remembers, "one by one researchers were dropping out of human retrovirology and by 1977 the field was getting pretty thin. Worse, those who had left did so convinced that they had wasted a lot of their time—that there was never going to be a human retrovirus."

In fact, it was also around this time that the Virus Cancer Program, having begun with such high expectations, came to an inglorious end. Scientifically, the problem was that no one could supply clear evidence of any kind of human tumor virus, not even a DNA virus, and most researchers refused to concede that viruses played any important role in human cancers. Politically, the Virus Cancer Program was vulnerable because it had attracted a great deal of money and attention and had failed to produce dramatic, visible results. When the charge was made that the program had been lax in overseeing its contract arrangements, an independent committee, headed by Norton Zinder of Rockefeller University, was brought in. Following the release of the committee's report, the program was canceled.

The Virus Cancer Program did not come and go without a legacy. Important work in a variety of fields, including the study of oncogenes, was first undertaken and supported during the seven years of its existence. Many important methods and reagents were developed by scientists funded under the Virus Cancer Program, methods and reagents that were extremely useful for the future, as we will soon see. Dick Rauscher, John Moloney, and Robert Huebner helped direct money to important yet previously unexplored areas of cancer research. But in another respect as well, the Virus Cancer Program left its special mark on biomedical research. Its demise and the attacks on its directors had a special effect on those of us in sister programs within NCI. The message was clear. In the absence of extremely forceful leadership of the NIH, of the type James Shannon was said to have provided, decisions about the direction and funding of biomedical research can suddenly become very political, and woe unto any scientist who finds himself in the field of fire.

Now, with Poiesz, here was a young scientist who needed to achieve something, investing his own future in the hope that he would be part of the discovery of the first human retrovirus. Joining with Ruscetti and others in the laboratory, Poiesz began efforts to grow human T-cell leukemias with IL (Interleukin)-2. It was already clear to us by this time that normal T-cells did not respond and grow with

IL-2 unless stimulated by PHA or certain antigens. We suspected and later reported suggestive evidence that this was due to the lack on "resting" T-cells of IL-2 receptors, which were formed on the surface of a T-cell only after it had been activated in response to a foreign antigen or to a stimulant (like PHA).

Surprisingly, we were able to get *leukemic* T-cells from some adult patients with T-cell leukemias to respond directly to IL-2 without the benefit of prior stimulation; we could grow them simply by adding IL-2. Later it would be unambiguously demonstrated by Warner Greene and Thomas Waldmann and their co-workers at NCI that that was because these leukemic T-cells abnormally and continuously express on their surface the receptor molecule that binds IL-2. And it would be from just such leukemias that shortly thereafter our group would find the first human retrovirus.

I have often been asked how we felt at the moment of discovery, whether we jumped up and shouted "Eureka!" The answer is no. In science, finding something and confirming the find are separate processes. With HL-23, we had done a number of confirmatory experiments prior to publishing our findings and had still been burned. No one in the lab had to remind the others that there was much to be done before we could be convinced that what we had found was indeed a retrovirus and not an artifact; that we had found a human retrovirus and not a contaminant animal virus; that we had found a pathogenic virus and not something irrelevant as, for example, I believed an endogenous retrovirus would be; and that our observation was reproducible. Reitz vividly remembers how we all felt at this time.

"Our emotions were tempered with a lot of paranoia because of the HL-23 debacle. It wasn't just Hershey—it was the whole period of a year or two where it affected everything anyone at the lab did. You had the feeling that people just wouldn't believe stuff coming out of the lab. Or anything that we would report. Because of HL-23 I just knew that any oversight regarding this new human retrovirus would be fatal to the lab. There would never be another chance."

It is now time to stop the discussion about developments in my laboratory. I will return to these events later. But first I need to describe some of the physical characteristics of a prototypical retrovirus and discuss its possible origins. Its envelope consists of protruding knob-like structures with two components: an outer part and a rod-like structure that digs into the interior of the physical virus particle, penetrating a bilayer of fatty material (the so-called lipid bilayer). Chemically the viral envelope is a glycoprotein, that is, a protein with several sugar-like components attached to it. The envelope is a very important structure. It is the principal initial factor in determining the kinds of cells a virus can infect, because it is the envelope that first attaches the virus to its target cell. We think it usually does this by binding its outer part to some cell membrane component, which we can now refer to as the receptor for the virus, though obviously this cell membrane component surely evolved to serve the cell—for example, as a homing device for interactions with other cells or a locking mechanism for receiving extracellular signals—and not to facilitate intrusion by the virus. In a sort of tit-for-tat arrangement, some antibodies have learned to use the viral envelope for blocking virus infectivity. Therefore, the envelope is generally the key component of any vaccine effort.

The envelopes of retroviruses are named according to their size. For example, the AIDS virus envelope has a molecular weight of 160,000. The outer component is 120,000, and the rod-like interior part about 41,000. They are termed gp (for glycoprotein) 160, gp120, and gp41, respectively.

The lipid bilayer traversed by the rod-like part of the envelope is acquired by the virus as it pinches off from the cell membrane during the final stages of the viral replication cycle. The virus interior is composed of the core, which contains the viral genetic information (the viral genome) housed within core proteins.* The viral genome consists of two identical strands of RNA. Closely associated with the RNA molecules are molecules of the enzyme reverse transcriptase (RT). The core is formed by a few relatively small proteins, also named by their size. For example, the major core protein (in terms of amount) for all human retroviruses has a molecular weight of about 24,000 and is called p24. There are also smaller core proteins, whose specific structural duties we do not need to understand in detail for our purposes here.

I have noted that the lipid bilayer is acquired by the virus from the cell, but the proteins of the virus—the core proteins, the reverse

*One core protein called p7, or nucleocapsid, is intimately associated with the viral RNA. Another, p24, forms the core "wall." Outside of it, that is, just beneath the envelope, is still another "shell." This is p17, also known as the matrix protein.

transcriptase, and the envelope—have to be made in each cycle of infection. They are products of viral genes. The vast majority of animal retroviruses have only three genes: *gag,* which codes for the core proteins; *pol,* coding a few enzymes such as RT; and *env,* which codes for the envelope. They are located in this order in the RNA genome. The human blood tests for the AIDS virus and for leukemia viruses aim to detect antibodies present in human blood that react specifically with these viral structural proteins.

The major steps in the replication cycle of animal retroviruses can be divided into two parts: first, those in the incoming process, ending with the integration of the DNA provirus and mostly driven by viral components; and second, the outgoing process, ending with the formation of the whole virus as it exits from the cell. The latter is highly dependent on the machinery of the host cell.

The first phase in the incoming process can be further subdivided. Once the viral outer envelope component binds to the receptor site on the cell membrane, there may be rapid fusion of the viral envelope to the cell membrane, a process that is still poorly understood. This leads to the penetration of the viral core into the cell, its transit to the cell nucleus, and at some stage viral DNA synthesis.

The DNA synthesis reaction is directed by the viral RNA acting as a template and is catalyzed by the enzyme RT. Probably the reaction is initiated by the availability of the correct deoxyribonucleotides, the building blocks for all DNA. Since they are available only in the cell, the viral components can make DNA only after entry into the cell.

After two strands of DNA are made, the DNA integrates into the host cell chromosomal DNA (where now it is called the provirus), by a mechanism known for its precision and efficiency. Flanking the viral genes in viral RNA are always regulatory sequences that are duplicated when the viral DNA is made. These sequences are known as long terminal repeat, or LTR, sequences. They are important to integration of the provirus, and they will govern subsequent steps in the viral replication cycle. It is also at the LTR ends that the cellular sequences are joined to the viral sequences during integration.

The second phase of viral replication begins with the reformation of viral RNA in the cell's nucleus from the integrated DNA provirus. This reaction is initiated by signals contained within the LTR sequences, and it requires cellular factors, especially the cell enzyme that catalyzes RNA synthesis (RNA polymerase II). Some of the newly formed viral RNA will be cut, portions removed, the remainder spliced, modified further, and transported to the cytoplasm, where it acts as a messenger RNA to direct synthesis of viral proteins, again completely dependent on the host cell for the housing and manufacturing tools.

Other RNAs will remain unspliced. Some of these full-length viral RNA molecules also act as messengers, but others will serve as new genomic RNA to be packaged in the viruses that will be formed. First they enter the cytoplasm and assemble with the newly made viral proteins near the cell membrane. There follows a poorly understood process of viral formation by budding-off of the cell membrane to complete the cycle of replication.*

Retroviruses are not thoroughbreds. When two or more infect the same cell, they often mix their proteins and sometimes even their genes, in several different ways. They may also mutate by making erroneous substitutions of some of their nucleotides, partly due to mistakes made by the enzyme RT during DNA synthesis. Sometimes they lose pieces of their genetic information. As mentioned, they occasionally acquire a sequence of DNA from the host cell. For a retrovirus to maintain its capacity to enter certain cells and to replicate—in short, to survive—it must, however, set limits on its tolerance of all these tendencies to change.

All retroviruses are very similar in their size, morphology, content, much of their genome, and replication cycle. But virologists have been able to subclassify them, based on their observations that, despite common characteristics, the biological effects of retroviruses are diverse. At first they were subcategorized only according to the different diseases they induced: leukemia viruses, sarcoma viruses, carcinoma viruses, and so on. Refinements in our understanding of their structure led to an electron microscopic subclassification: type-A, so far restricted to mice; type-B, so far restricted to mice and rats; type-C, by far the most common, present in many species and including the vast majority of those that cause cancer as well as most of the rampant endogenous retroviruses; and type-D, found mostly in some monkeys

*As much more is known about the genetics and replication cycle of retroviruses than I can discuss in the scope of this chapter, the interested reader is referred, in particular, to Howard Varmus's review of it in *Science* (1989) and to the October 1988 issue of *Scientific American* for more details.

and associated with cancer and other diseases, but not yet established as the cause of any.

Almost forgotten until recently was another subfamily of retroviruses which were known only in certain ungulates (sheep, goats, horses, and cows) and were the province almost solely of a few specialized veterinarians. Called lentiretroviruses (from the Latin word for "slow"—perhaps inappropriately, because most other retroviruses cause disease just as slowly or even more slowly than these), they induce mostly chronic degenerative and neurological diseases. Work on the human AIDS virus, which belongs to this subgroup, stimulated very recent discoveries of other lentiretroviruses in several other species, especially among monkeys.

Still other classifications of retroviruses have been based on laboratory cell-culture effects. Rigid subclassification has its problems, though. For instance, we have learned that leukemia viruses will at times cause nonmalignant disease. The cat leukemia virus can cause anemia, immune deficiency, abortion, and several different forms of leukemia. The type of disease caused can turn on very minor differences in the virus's genetic makeup. Differences in the virus makeup that correlate with differences in the disease it causes are chiefly found in the envelope, but we usually do not know why.

Thanks to new technology, scientists have also developed new criteria for classification. The development of immunological and molecular techniques has allowed us to relate one virus to another with much more precision. Most of the time this has substantiated and enhanced the earlier classifications, but it has at other times forced a rethinking. Most retroviruses with type-C morphology are oncoviruses (cancer-causing), and most of those with the lentiretrovirus type of morphology are in another related group. These are now the two most important groups of retroviruses and *the only ones known to be present in humans.* But non–type-C retroviruses can produce cancer in at least some animals, and sometimes a strain of a type-C retrovirus (and probably some strains of lentiretrovirus as well) produces an unexpected type of disease or no disease at all. Certainly we no longer think within the narrow confines of the one microbe/one disease idea implicit in Robert Koch's postulates (see chapter 15).

Earlier I discussed how, like any *bona fide* virus, retroviruses can be transmitted as extracellular particles, sometimes also within infected cells, as may occur when there is blood contamination from one person to another. I contrasted these exogenous viruses to those endogenous retroviruses that are transmitted genetically in the germ cells (sperm and egg) in the form of DNA proviruses. I have noted that endogenous proviruses are ubiquitous: part or all of a provirus complement has been present in every vertebrate species where we have looked for them. It appears now that the LTR components of the proviruses are related to a large class of genetic elements that can change from one position to another in normal cell DNA.*

These elements are called *retrotransposons.* They were first identified in lower forms of life such as yeast and are reminiscent of *transposons*—genetic elements that change position—which were first discovered in experiments with corn. Barbara McClintock received the Nobel Prize for this work.

These moving bits of DNA can have an important biological function, influencing the behavior of this or that gene of a cell, depending on their location at a given period. Perhaps retroviruses evolved from retrotransposons, becoming more complex. A more appealing possibility may be that retroviruses evolved with living forms almost from the start and, after becoming endogenous proviruses, often degenerated in time to become incomplete proviruses and retrotransposons. Conversely, an endogenous retrovirus with full coding capacity to produce a virus may do so with the likelihood of eventually yielding variants that become infectious for the same animal species, or even escaping into a new species, now as an exogenous retrovirus. As I discussed earlier, this may lead to disease, particularly when the recipient species has not previously experienced an infection with this kind of retrovirus. I wonder whether there is a cyclical process in its evolution: a sequence of exogenous virus → disease → adaptation → germ line infection → endogenous virus → defective endogenous virus → retrotransposons, followed by the reemergence of a fully formed endogenous virus from retrotransposons and eventually back again to the infectious viruses.

*Important portions of the makeup (nucleotide sequences) of viroids (see chapter 3) are similar to these retrovirus sequences. Some believe viroids are degenerative forms of ancient retrovirus proviruses.

6

Discovery of a Cancer Virus: The First Human Retrovirus

It took over a year of careful work to confirm that we had found a human retrovirus—a process that involved several members of my laboratory. The original detection of HTLV-1 had been based solely on positive reverse transcriptase–like activity in the fluid of a leukemic T-cell line established from a patient with a lymphoma of the T helper cell, the T4 lymphocyte.*

These original reverse transcriptase assays had to be redone with careful attention to the assay conditions, to distinguish the enzyme from normal cellular DNA polymerases. It was decided that Poiesz would do this, but to gather more convincing evidence, I suggested he also use antibodies—made in our laboratory by several former postdocs over a period of several years, each antibody specific for reverse transcriptase of a different animal retrovirus. Using them showed that the enzyme from the human leukemic cells was not from one of the animal retroviruses. Earlier, others in my laboratory, especially Marjorie Robert-Guroff, had also developed specific antibodies made to the normal human DNA polymerases. These reagents were

*Because patients with such a lymphoma often also develop a high white-blood-cell count, we often call it a leukemia. In animal retrovirology the viruses said to cause leukemia often produce what clinicians would call a lymphoma. Even though such lymphomas are often associated with a "spillover" of tumor cells into the blood, some hematologists like to distinguish them from true leukemias, which have a clear-cut origin in the bone marrow. In contrast, lymphomas are believed to originate often outside the bone marrow and blood—for example, in lymph nodes.

also used by Poiesz, and they too showed minimal or no cross-reactivity with the reverse transcriptase released into the fluid by cultured leukemia T-cells.

We could now be reasonably sure of having a novel reverse transcriptase. Yet, it would be wise to know the properties of the reverse transcriptase in its purified state, and this called for someone with more experience in enzymology. H. Rho, who joined the lab in 1978 and now works in his native Korea, purified the reverse transcriptase and showed biochemically that the purified enzyme had all the characteristics of a viral enzyme. Further, he did more thorough testing of this reverse transcriptase with several different antibodies made earlier against reverse transcriptase from different classes of animal retroviruses. He, too, showed this reverse transcriptase to be distinct from any of the earlier animal ones.

Meanwhile, Poiesz alone could not prove that the particle in which the reverse transcriptase was found was really a virus. That would require: (1) electron microscopy to show structure, and (2) demonstration of biological activity such as transmissibility. So we sent out our samples for electron microscopy through the collaborative arrangements I had made some time before with another group. Not surprisingly, the first electron micrographs failed to detect retroviruses, but eventually we were able to get pictures that convincingly showed the presence of virus particles identified as type-C–like retroviruses. Despite the fact that the evidence was now convincing and that virus particles could be seen, I still had doubts. One of Poiesz's key contributions was establishing close collaborative contacts with a clinical group headed by a clinical scientist named John Minna. Though we had had many previous clinical collaborations, this one was with a relatively newly formed NCI group, one of whose interests at that time was the therapy of leukemias and lymphomas of T-cells. Within this group a cell line was established from a patient with T-cell leukemia. Poiesz had his positive results with the fluid from the cell line. We had clearly proved that retrovirus particles were present, and our reverse transcriptase data strongly indicated that the particles were not from one of the well-known animal retroviruses.

But now I started to worry about the nutrient broth we had used to grow the cells: it might be the source of a contaminating animal virus. Such broths generally contain media, which is a commercially available fluid containing measured amounts of numerous minerals, vitamins, and other nutrients. The other component is animal serum, usually from a calf. Could the calf serum used by us have come from a calf infected with the bovine leukemia retrovirus? No one had dealt with this possibility.

Only one person I knew could give me the necessary tools (reagents) to answer this question: Arsene Burny. I had first met Arsene in 1970, when he was among the brightest and the best of Sol Spiegelman's group. In the intervening years he had become an expert on bovine leukemia virus in his native Belgium and a close friend. Without a minute's thought to the time—it was 10 P.M. Brussels time—or to the fact that it was Christmas Day, I rang him up on the telephone.

"I need your help, Arsene," I told him. "I think we've found the first human retrovirus and we've ruled out conventional animal viruses, but we used fetal calf serum as our nutrient broth. What if it was contaminated with bovine leukemia retrovirus?"

To this day, Arsene regales his lecture audiences with this story. Of course, he immediately sent me the reagents. And, of course, our isolation did not turn out to be bovine leukemia retrovirus. But before we hung up he asked me whether I realized what time and what day it was and then said, "Only the two of us would be crazy enough to be working Christmas Day!"

We had now excluded possible contamination with all common laboratory retroviruses as well as possible contamination from calf serum. But what I still fretted about was contamination by a less well known retrovirus or an unknown one, an animal retrovirus for which there were no specific reagents, such as our specific antibodies to reverse transcriptase. I approached this problem in two ways. First, I required that we also find the virus in the *fresh* blood of our original patient. I knew that a second NCI colleague, the immunologist Tom Waldmann, had seen and studied the patient. He immediately supplied us with specimens obtained directly from the patient, not yet cultured. Poiesz was easily able to detect the telltale sign, the positive reverse transcriptase activity, in particles present in the fluid of the T4 lymphocytes he had now newly cultured with IL-2.

Important culturing work was also carried out in my group by Michiyuki Maeda, who came to LTCB as a visiting scientist in 1981 to study some aspects of the interaction of our new virus with T-cells (and is now with Kyoto University in Japan). Michi made many single-cell clones of virus-infected cells and expanded each in cell culture.

This helped in many of the subsequent studies of my staff that documented that this was indeed a human retrovirus.

Further evidence would come from another direction. As I have noted, the core of any retrovirus is made up of several proteins, and one such protein (called the major core protein) is particularly well defined. Among all the then-known animal retroviruses of type-C morphology, these proteins had some relatedness to one another. But there are techniques to distinguish them. I asked V. Kalyanaraman (known as "Kaly" in the lab) and Sarang to focus on the analyses of this protein from the candidate human retrovirus.

Kaly first joined us as a postdoctoral Fellow in 1974. Like Sarngadharan, he came from southern India, and also like Sarngadharan he enjoyed and excelled at protein chemistry. They worked together. In a rather short time they purified this protein and tested it with various antibodies (as in the case of reverse transcriptase). Later they elucidated its chemical structure (order of amino acids), proving that the virus was novel by these criteria.

The most unambiguous test to end these concerns forever would be to demonstrate that present in the blood of the patient from whom the virus was believed to be isolated were specific antibodies to the virus; this would prove that the patient had been infected by the virus. A second, equally clear result would be to find the viral genes integrated in the DNA of his fresh leukemic blood cells. Marjorie Robert-Guroff (working with a new postdoc Fellow, Larry Posner) and, independently, Kaly, and later Carl Saxinger, each approached the task using a different technique, all demonstrating that this patient's sera had antibodies in it that specifically reacted to the proteins of the virus. Simultaneous work came from Marv Reitz and subsequently from Flossie Wong-Staal. They did the molecular analyses of the viral genes. One such study (by Reitz) showed that nucleotide sequences of the virus were present in the DNA of the patient's cells. This and the serum antibody tests definitively established that this was not a contaminant but a retrovirus that really had infected this individual.

The virus could be continuously propagated in the cell line from this patient. This allowed us to collect rather large amounts of viral proteins—enough for Robert-Guroff, Posner, and Kaly to test sera from large numbers of people (healthy people as well as those with leukemias and lymphomas) from different parts of the world. Although we would soon make additional isolates of this virus, clearly if the serum tests could be made sensitive and specific (they could), there would really be no need to keep re-isolating the virus with all the cost in time and money and the danger of contamination. Populations could now be screened by a simple, inexpensive test. Such sero-epidemiological studies brought us into almost daily contact with the NCI epidemiologist William Blattner, who would be a key collaborator in our work from then on. These serum antibody tests, developed by us in 1980, would also become the basis for the blood test for this leukemia virus, which would be put into effect in U.S. blood banks years later (in 1988). It would also be the basic forerunner of the 1984 blood test for the AIDS virus which, of course, is another story.

No matter how refined the results, however, a particle cannot be called a virus unless it can be shown to have biological activity, such as its transmission (while in the laboratory) into an animal or its passage from one cell to another new target cell or some gross effect on some target cell. Although Poiesz had detected chronic production of reverse transcriptase activity in a cell line he had acquired from Minna's laboratory, and now also short-term production with fresh primary blood cells newly taken from the patient and fed by IL-2, he had not demonstrated any infectivity or other biological effects of the virus. (This would not come until late 1981/early 1982, with the work of newcomer Mikulas Popovic and of Markham, Salahuddin, and a Danish visiting scientist, Gunhild Vejlsard, now at the University of Copenhagen.) But we had much *indirect* evidence for its biological activity, in the demonstration of antibodies in the patient's sera (by Kaly, Posner, Robert-Guroff, and Saxinger) and of viral genes in the DNA of leukemic cells (by Reitz). This clear-cut evidence showed, albeit indirectly, that the patient was infected; thus, this was indeed an infectious exogenous virus.

Reitz describes his reaction as his end of the process was completed: "When I was in graduate school, I used to go caving. One time we went into a cave where we were looking for an unexplored passageway. There was this room that was not very high, with hallways going off in about three different directions. We took two of the directions and they just dead-ended and people were wanting to leave and I said 'No, no, no, let me just check this last one out,' and I crawled a little way down and there was this pile of rocks, and I looked over the pile and it just went out into nothing. In some way you could tell when you got past the rocks that nobody had ever been there before. There were no footprints, no graffiti, no nothing, and it was kind of amazing in a way, an amazing feeling. And I yelled back to the other people to come on, but they thought I was trying to get them to come down a miserable crawlway, because cavers do that a lot. But I was jumping up and down so that they could hear that I was in a passage that had standing space, and as each one of them came crawling over the rock pile you could see their faces and you knew they all had the same feeling. Well, you know, I had a feeling similar to that as gradually the realization sunk in that we had really found the first human retrovirus."

My own reaction was tempered by the past. I wanted to see the virus independently isolated from another patient. Even though I knew we were right, I wanted to know it was something that could be done again by others following our protocols. Soon we obtained a second isolate. Shortly after Poiesz left the laboratory to go to the State University of New York Medical Center in Syracuse, Popovic joined us and made many more isolates in rather quick succession, as did Markham and Salahuddin. Our preliminary antibody testing was also completely consistent with a role of this virus in the cause of these leukemias.

After almost all of this information was available to me, after it was discussed and rediscussed, then and only then did I feel it was time to state our results publicly, to allow publication of the first paper, and to conclude that we now had a human retrovirus. My plan was simple. We had to convince the critics as soon as possible. Four or five papers would be written for publication, submitted one after the other and intended to answer every possible significant question we could anticipate. I decided Poiesz could present some of the results at scientific meetings in 1979. Most responded with cautious interest, some with snickers. I submitted the first paper in mid-1980 to the *Proceedings of the U.S. National Academy of Science*, where I felt it would be assured of publication. A member of the academy can usually sponsor publication with success. I would not become a member until 1988 but the paper was sponsored by the late Henry Kaplan and published in December 1980.*

*Henry Kaplan of Stanford University was one of the greatest cancer biologists in the United States. In addition to his basic research on mouse leukemia viruses and on cell biology, he, more than anyone, was responsible for the radiation treatment of Hodgkin's disease and some other lymphomas that sometimes led to their cures.

I knew it would never swim alone. Other papers were sent to important virology and immunology specialty journals almost at the same time. The key one concerned the core protein by Sarngadharan and Kaly and was met with an uncharacteristically harsh reply. The journal editor rejected our submission, writing to us that he had no choice but to do so. He explained that our manuscript could not be considered in any form as there was nothing to be gained from resuming the controversy it stirred up over the existence of a human retrovirus.

"We were shocked," recalls Sarngadharan. "What a reward for breaking in the new era of human retrovirology. It took six months and a lot of discussions, but what was rejected outright was later accepted back."

After being refused for publication in 1980, on March 6, 1981, the article was finally accepted for publication in this same journal. By 1981 we had, in fact, published several papers and, as noted, already re-isolated the virus from other patients. We named it Human T-Cell Leukemia Virus (HTLV), in accordance with long-established principles of nomenclature for animal retroviruses (feline leukemia virus, murine leukemia virus, gibbon ape leukemia virus, bovine leukemia virus, and so on). We specified T-cell because each positive case involved T-cells and almost always a T-cell subset known as the T4 or $CD4^+$ T-cell. Because later we learned that the virus also can cause other forms of disease, we and others later suggested that *lymphotropic* might be a better term than *leukemia.* But I was not at first convinced that the change was appropriate, because most animal retroviruses that cause leukemia can also cause other diseases. Both terms are used today.

Our first isolate of HTLV had been made from the T-cells of a young black man from Alabama who was under study at the National Cancer Institute by clinical and laboratory investigators. This man had an aggressive T4-cell malignancy. He was a heterosexual with no relevant medical or personal history. We had obtained two more independent isolates from this patient. Our next isolate was from a middle-aged black woman from the Caribbean; the third contributor was a middle-aged white merchant marine with a history of extensive travel to Japan and the Caribbean and encounters with prostitutes. More cases from the Caribbean region would follow. All the cases would have some common clinical features: the malignant cells were the $CD4^+$ T-cell (T4-cell), the disease usually spread rapidly, and skin abnormalities were common, especially nodular lesions due to infiltration of the skin with the leukemic cells.

Meanwhile, our collaborator, the outstanding immunologist Thomas Waldmann of the National Cancer Institute, told us about the work of a new Japanese Fellow in his laboratory. H. Uchiyama had worked with Professor Kiyoshi Takatsuki in epidemiological studies of Japanese patients with leukemia. In 1977 they had reported finding clusters of T-cell (particularly T4-cell) leukemias in southwest Japan and especially in the islands of Kyushu, Shikoku, and Okinawa. They called this tumor adult T-cell leukemia/lymphomas, or ATL. A major convergence between our respective groups would now occur.

Before Takatsuki's reports, epidemiological studies of leukemias in Japan suggested that they were not much different in number and type from those that had occurred in the Americas, Europe, and most other places, with the exception of the transient increase in some forms of leukemias in the regions of Nagasaki and Hiroshima due to radiation from the atomic explosions. But epidemiological studies should be reevaluated when technology improves, and this is no better illustrated anywhere in medicine than in this story.

In earlier studies, T-cells and B-cells had not been distinguished, and certainly subtypes of T-cells could not be because they were then unknown. Leukemias of the lymphocyte were simply lumped as acute or chronic (according to the degree of aggressiveness) lymphocytic leukemias. Takatsuki reexamined cases of leukemia after specific reagents and assays became available for their more precise subclassification. His data quickly revealed an increase in T-cell leukemias in the southern portions of Japan that was made more dramatic when the leukemias were further distinguished as T4, T8, or more immature cell types.* The Japanese cases were usually aggressive and often affected the skin. The similarities to our few HTLV-positive cases in the United States were obvious, even though the name of the disease was different and no cluster was apparent in the United States.

Around the summer of 1980, I initiated a collaboration with a longtime friend, the late Yohei Ito, who was chairman of microbiology and dean of the medical school at Kyoto University and a

*T-cells originate from a primitive and probably nonfunctional bone marrow cell which migrates to the thymus gland where steps in the maturation to functional cells occur. The more mature cells exit the thymus and can conveniently be found in blood.

virologist with years of experience with two kinds of DNA viruses: EBV and certain papilloma viruses. By late 1980/early 1981 we found that eight out of eight Japanese cases of this type had serum antibodies against HTLV. Also, in a small survey we found positive results indicative of infection in about 3 to 10 percent of "healthy" people from these southern islands and very little evidence of infection elsewhere in Japan, where ATL was low or nonexistent according to Takatsuki's results. We now had evidence for clusters of HTLV infection exactly where clusters of ATL occurred and 100 percent linkage to disease; we also now knew that retroviruses in animals often caused leukemia. These results strongly argued for the direct role of HTLV in these cancers.

In early March 1981 a small meeting was organized by Professor Ito in the beautiful mountains by Lake Miwa, outside of Kyoto, in order to facilitate major collaborative efforts between us and others in Japan, and for us to give Japanese investigators some important reagents, such as serum from animals containing specific antibodies to HTLV we had induced by injecting them a month earlier with HTLV particles or proteins purified from HTLV as well as virus-producing cell lines. The idea was to withhold publication of the results on the Japanese patients until a more extensive study was carried out with many groups of Japanese investigators. I had already developed close collaborative arrangements with William Blattner and his epidemiological group at the National Cancer Institute, particularly for serum studies of people in the Western hemisphere. Blattner, an M.D. with public health and epidemiological interests, keen insight, and great dedication to his work, was a vital force in these and in virtually all subsequent epidemiological studies designed to link or eliminate viruses isolated in my laboratory to various human diseases, but he was not involved in these Japanese studies or with me during this trip to Japan. A few other co-workers and collaborators accompanied me on the eventful trip to Lake Miwa.

The meeting consisted mostly of a series of long reports by my colleagues and myself about our already published work on HTLV and a discussion of our recent, unpublished results on samples from Japan. A few Japanese pathologists and epidemiologists chiefly discussed some chronic parasitic diseases as a possible causative factor of ATL, but we learned a great deal from them about the pathology, clinical features, and epidemiology of this leukemia. Since no parasite was known that had ever caused leukemia in animals and in humans (and conceptually this was not an attractive idea), and since retroviruses were a common cause of leukemia in animals, I felt the notion of those inclined to a parasitic cause to be misguided, unless the parasitic infestation was a co-factor, something that perhaps might accelerate the pace of virus replication and spread by fostering activation of lymphocytes.

At the end of the meeting, however, Yorio Hinuma of Kyoto University announced a change in the title of his talk. Originally scheduled to speak on his herpes virus research, his career-long specialty, he announced that he too had found a new retrovirus, which had been isolated from a Japanese ATL case. He and his collaborators also had preliminary results of antibody data in patient's sera similar to ours, though they were based on a simple but less accurate test, an immunofluorescent assay. He was not ready to collaborate. He had not yet published anything. Instead, a few weeks later he presented his results at a press conference in Japan where he announced isolation of the first human leukemia virus and the first direct cancer-causing virus in humans. He named the virus ATLV (adult T-cell leukemia virus). But it was likely that this virus was the same as HTLV.

Later a cooperative study with several Japanese scientists, headed by Mitsuyaki Yoshida of the Tokyo Cancer Institute, was established and showed that HTLV and ATLV were the same kind of virus. Japanese investigators, our group, and representatives from England agreed that the virus would be called HTLV, that the disease would be called ATL, and that more collaboration was imperative.

The isolate of HTLV by the Japanese group was published in June 1981. It was important for many reasons. First, it provided useful, perhaps needed, confirmation of our work. Second, it linked the virus to a cluster of diseases. Third, we learned that the T-cells from which the virus was found in Japan had been established in tissue culture without IL-2! Isao Miyoshi, then a temporary collaborator of Hinuma, made their key observation.

Miyoshi had established the cells producing their virus and growing in permanent culture by co-cultivating ATL cells with normal cells. His goal: to use the normal human cells as a source of some hoped-for growth factor with the intent of growing the ATL cells. Unknown to him, HTLV was present in the ATL cells; it transferred to the normal cells and converted them into permanently growing T-cell lines.

For reasons we still do not understand, "normal" cells newly infected with HTLV will usually "overgrow" the ATL cells as the newly infected cells become immortalized by the virus, as they did in this case. This cell line was the one used by Hinuma, Miyoshi, and Yoshida in their identification of HTLV in Japan, but Miyoshi's experiment also suggested that HTLV was a transforming virus, or at least it conveyed one of the steps leading to cancer cell immortalization. It was, then, a virus that could be used in the laboratory to study at least one step of the cellular and molecular mechanisms of a viral-induced cancer of humans.

That this laboratory result—immortalization of cells—was very likely relevant to the disease was emphasized by the fact that the targets of HTLV in these experiments were T lymphocytes, and usually it was the T4 lymphocytes that became the dominant immortalized cells in the cell culture system. This was precisely the cell type that usually became leukemic when associated with HTLV infection in the patient. Consequently, these findings also boosted our confidence in thinking HTLV also caused the leukemia.

Czechoslovakia has contributed to science during this century more than its size, population, and, in particular, its political difficulties would predict, and virology has been one of the areas in which it has made special contributions. Among its leaders in virology was Jan Svoboda of Prague, who was among the few who considered and helped further Temin's provirus hypothesis in the 1960s. Some of his other contributions helped us in our thinking and techniques for transmission of viruses to cell lines. Svoboda's impact on virology was impaired in 1968 with the violent overthrow of the Dubček government. But he could still be an active voice in a small laboratory, and in the 1970s he had the advantage of being far removed from the pressures against working or thinking about human retroviruses. In fact, some members of his "school" still thought endeavoring to find them a worthwhile pursuit. I met one of these people in a pub in Norway in the mid-1970s.

That evening over a Slivowitz brandy, Mikulas Popovic told me of his interests. Sooner than either of us could have expected, Popovic joined our laboratory. Shortly afterward, he took the HTLV system and Miyoshi's transformation results and made the *in vitro* transformation (immortalization) of T-cells a precise, routine laboratory system.

Exploiting the many reagents to HTLV now available in our laboratory and our extensive clinical collaborations, he also was able to make many more independent isolates of the virus. He did this by combining our IL-2 approach (growing primary leukemic T-cells, usually obtained from blood, with IL-2) and, after HTLV was detected in this way, transmitting it to target T4-cells present in human umbilical-cord blood. These then become immortalized and usually the cell line so formed will continuously produce HTLV.* In other words, he more or less combined our older approach with methods that came out of Miyoshi's experiments. The field of human retrovirology now rapidly expanded. Scientists from most European countries, many from Japan and the United States, and a few in other parts of the Americas, began studies at the molecular, cellular, clinical, and epidemiological levels.

During a short trip to Europe in 1981, William Blattner and I were asked by the then elderly and well-known hematologist Sir John Dacie to take part in a meeting, unusual for its small size, conciseness of subject, and subsequent impact. Joining us were a few clinical hematologists, including Daniel Catovsky of the Hammersmith Hospital, who, like Sir John Dacie, were interested in leukemias and their origin; a few virologists—William and Oswald Jarrett, Robin Weiss, and Tony Epstein; and a cell biologist with expertise in leukemic cells, Mel Greaves. Within the first hour, Catovsky revealed that he had a collection of eight or nine patients with signs and symptoms similar to the ATL cases in Japan and the sporadic HTLV-positive T-cell malignancy cases we had found in the United States. This was interesting, but what was important was their origin: they were all immigrants or descendants of immigrants from various Caribbean islands.

This was one of the few moments I have experienced in science where a hasty conclusion would turn out to be fully correct. We knew in an instant that they would be HTLV-positive (all were), and this explained our too frequent finding of this disease and virus among African Americans and West Indians in the United States. The Carib-

*I do not know why T4-cells and other T-cells from the blood of a baby's umbilical cord are more susceptible to being immortalized by HTLV than are T-cells from adult blood. The possibilities that come to mind include that the real target is a more immature T-cell found in greater numbers in the blood of a newborn than in an adult, which after growth expresses the T4 marker (CD4 molecule) on its surface. Another is the presence of a more advanced immune system in adults that can block growth of the infected cell much more successfully than can immune cells from an infant.

bean was an endemic area! I thought quickly about the scientific and epidemiological implications.

Blattner moved to establish links throughout the Caribbean basin—Jamaica, Trinidad, and Panama would be particular points of our focus—and in the next few years brought in much information on the clinical-epidemiological features of HTLV disease in the Americas. Along with results from Japanese investigators, we would learn that the virus spread by blood and sexual contact, as well as from mother to offspring. In heterosexual contact, the virus was effectively transmitted from male to female via semen; female-to-male transmission occurred less frequently. Males were more frequently infected by their mothers' milk and possibly also by *in utero* infection.

The disease also had a variable latency—that is, the time from infection to the time of disease development varied: it could occur in a few years; more often it took decades. A baby infected by his or her mother would usually not develop leukemia until age thirty or forty! Yet, we found cases where leukemia developed in a person from a nonendemic area apparently only a few years after sexual encounters in an HTLV-endemic region. I wondered whether this more rapid pace was due to greater susceptibility of a viral-virgin population. It was about this time that a young postdoc from Zurich, Jörg Schüpbach, joined our lab. He would become very important to our later work in the breakthrough development of the AIDS blood test. My associates Carl Saxinger, Kaly, and Marjorie Robert-Guroff did most of the assays for antibodies in sera selected by Blattner and his colleagues from our clinical collaborators, especially Nigel Gibbs in Jamaica and Courtenay Bartholomew in Trinidad. Later studies of Guy de Thé from Lyon proved that while some Caribbean islands had significant infection (2 to 5 percent of the population), as we had already shown with a narrower study, other islands had no HTLV at all.

The studies were expanded. In the United States the virus was scattered here and there, chiefly but not solely among black populations and in the southeast. Bart Haynes, Dani Bolognesi, and their co-workers at Duke University were the first in the United States to confirm and extend our results. They reported successful isolation of an HTLV-1 from a woman with ATL. She lived in North Carolina, but she had been born in southern Japan.

Tests on native European populations were for the most part negative. A report from Amsterdam confirmed our experience with Catovsky. A cluster of HTLV leukemia was found in West Indian immigrants to Holland. A low rate of infection was found in the native population of Spain; and a strange and still unexplained cluster of infection was identified by Italian workers (Vittorio Manzari and, independently, Olivier Varnier and their colleagues) in the native population of a small region of southeastern Italy (the area of Lecce in the region of Apulia); and most recently, others have found that select populations in Iran and Egypt, especially those of Jewish origin, have endemic infection. More important, we and others found sub-Saharan Africa to be a particularly endemic region. Finally, studies of immigrants from the West Indies to Europe also strongly suggested that no environmental co-factor was needed for the leukemia to develop—only the virus. Adults infected as babies in an endemic area and remaining in the endemic semitropical region seemed to have the same incidence of leukemia development as those who moved as children to Europe.

Our knowledge about the relationship between the virus and the disease also expanded. It became clear that no more than 50 to 70 percent of HTLV-1–positive T-cell leukemias were the typical ATL cases described by Takatsuki: an acute disease with frequent skin involvement, leukemic T4-cells in the blood, abnormal convoluted nuclei of these cells, and often a high blood calcium. Other cases were more chronic, and although still involving the T4-cell, the leukemic cell did not have the convoluted nucleus or "spill" into the bloodstream. Rather, these HTLV cases were more like what is called a lymphoma with tissue and lymph node involvement only, taking a slower clinical course. Some epidemiologists also obtained estimates of the occurrence of disease following infection. The results have varied according to different regions, possibly reflecting differences in genetic susceptibility. Most suggest that roughly 1 percent of infected people develop leukemia after HTLV infection during their lifetimes, although more recent results suggest that the true number may be higher—3 to 4 percent.

Two other important clinical leads came from the epidemiological results and insights of others. By 1981 I had discussed the developments of the HTLV story with my longtime colleague and friend at Harvard, Max Essex, and he now decided to leave his feline leukemia virus work to focus on human retroviruses. One of his early contributions was his finding that HTLV-positive people (in a Japanese study) had a higher incidence of infectious disease. This report was supported by Takatsuki's findings that individuals with ATL developed opportunistic infections with greater frequency and severity than occurred in other cancers and that they often occurred before the cancer became severe.* In other words, patients with ATL appeared immune-suppressed in some specific but poorly defined manner that was beyond a level that might be considered to be secondary to the cancer. It implied that HTLV was immune-suppressive.

These results stimulated collaborative experiments between my laboratory and immunologists at Columbia University (especially Nicole Suciu-Foca) and by Anthony Fauci and his group at NIH. These experiments showed that HTLV could indeed harm the immune function of the critical T4-cell. A second clinical lead from France was much more surprising.

It is not easy for a scientist to have one foot in France and one in the United States. Guy de Thé seems to have succeeded with feet to spare. He also managed to be associated with a vineyard in Burgundy; with good cooking in Lyon; with epidemiology in Africa, the Caribbean, and China; and with both DNA and RNA tumor viruses. Originally, he was instrumental in the African studies of the 1970s that linked infection with the herpes virus EBV, coupled with malarial infestation, to an increased risk of developing Burkitt's lymphoma (a cancer named for Dennis Burkitt, the Irish surgeon who first described this tumor). Subsequently, he helped Chinese workers do the epidemiological studies of the possible role of EBV and some chemical carcinogens in cancer of the nasopharynx which occurs with high frequency in parts of China. He was ready to do something in the epidemiology of retroviruses—and he did.

I already mentioned his finding of HTLV presence in high amounts in some Caribbean islands and its total absence in others. Following the beginning studies of Blattner's group and mine indicating that HTLV was endemic in sub-Saharan Africa, de Thé showed that it was strikingly regional, varying from tribe to tribe. We had also learned

*Opportunistic infections indicate disease due to a microbe that is ordinarily not pathogenic (disease-causing). In immune impairment, some nonpathogenic microbes can cause disease.

from Japanese studies and from our own that HTLV tended to be transmitted within families and to stay within families for generations. Thus, it was likely that the variation in the Caribbean islands was dependent on the exact African origin of the people.

But later during these African studies, he and his young associate Antoine Gessain (now a postdoc with me in the LTCB) also noted a few patients with a neurological disorder called tropical spastic paraparesis (TSP), and published a slim but important study. Although received with considerable doubt, it was soon confirmed and extended, and HTLV is now the accepted cause of this degenerative disease of the spinal cord and brain (later specifically shown to be HTLV-1, as we will soon distinguish it from a second human retrovirus). The disease mimics multiple sclerosis in several respects, but unlike the waxing and waning in the clinical aspects of MS, the HTLV neurological disease is progressive. Some workers have recently suggested that the virus may cause neurological disease with greater frequency than it causes leukemia. Others, particularly Carleton Gajdusek and his longtime co-worker, C. Joseph Gibbs, have indicated that the frequency of HTLV neurological disease relative to HTLV leukemia varies according to the region.

This suggests that there could be a genetic factor predisposing an individual to either the neurological disease or the leukemia. Because no one has found that HTLV infects any cell of the brain or spinal cord and people with the HTLV neurological disease have higher levels of antibodies in their blood against HTLV than do other infected people, it could be that a genetic factor in such people leads to a diminished capacity to handle virus replication (the more antibodies probably means more virus) and that the disease comes about from autoimmunity. In other words, in attacking the virus the immune system overshoots and also interacts (destructively) with certain cells of the spinal cord and brain.

There are other possible mechanisms under study. For example, cells of the nervous system that are uninfected by HTLV could still be affected by proteins of degrading virus (or released by virus-infected cells) or by lymphokines released by infected T-cells. The HTLV that causes neurological disease may vary slightly from the one causing leukemia, as some slight differences have been found with a given animal retrovirus that causes different diseases. But increasing evidence now makes this possibility unlikely. For example, there are now several cases in which an infected person developed both ATL and the neurological disease.

These are important openings. On the basis of some preliminary data, it has even been suggested that multiple sclerosis itself may involve an HTLV-related virus. But this connection has so far eluded proof. Whether true or not, the clear link of HTLV-1 to a neurological disease similar to multiple sclerosis opened a door to the study of the mechanisms that lead to this kind of disease—the first opening of this kind we have had in humans.

During the past year, a meeting was held for the tenth anniversary of the discovery of HTLV. Surprising new information concerned the results of clinical investigators and epidemiologists whose findings indicate that HTLV may promote and perhaps even directly cause several other disorders.* These include some other neurological diseases, a rheumatoid arthritis–like disease, and some muscle inflammations.

*More recently, Mark Kaplan and his associates at Cornell University and the North Shore Hospital on Long Island have described a linkage of HTLV to a very serious infectious dermatitis of babies. Kaplan, who has been of immeasurable help to us in our virological studies over the years, not only with supplies of samples but with his extraordinary insights into human disease, also played his own role in future studies of human retroviruses, as I will mention later.

7

Discovery of the Second Human Retrovirus (and How the HTLVs Produce Disease)

In the beautiful spring days of 1981 in Venice, I attended a conference on blood cells. One very productive cell biologist, David Golde of the University of California in Los Angeles School of Medicine, was describing the properties of a new, permanently growing cell line he had established from cells of a patient with a rare leukemia. Called hairy-cell leukemia because of the hairlike filaments that extrude from the cell, it is usually of the lymphocytic B-cell type. Peculiarly, this case involved T4-cells. Golde named the cell line MO (using the patient's initials) and observed that it was like a little factory, constantly producing some growth factors and other biologically active molecules, the lymphokines I have already touched upon. In the discussion that followed his presentation, I noted that T4-cells transformed by HTLV (whether from fresh leukemic cells or "normal" T4-cells infected in the laboratory) similarly produced such factors, almost routinely. So perhaps that was not the most interesting thing about the MO cell line.

Since established T4-cell lines in animals and in humans had so far almost always been associated with a retrovirus, and since the type of leukemia of MO differed from any HTLV leukemia to date, I wondered aloud in the discussion of Golde's report whether an even more important use of the MO cells was as a source of another human retrovirus—one yet to be discovered. Probably because the HTLV data was not then completely known to all, my suggestion was not enthusiastically accepted. But some months later Golde sent me the fluid from his MO cell line to test for virus. (Golde had patent rights agreements with others, which probably prevented him from making the cells themselves widely available.)

HTLVs are isolated from fluid (as opposed to tissue) with difficulty. But he sent us substantial amounts of good-quality material. And with a dash of good luck, we succeeded in isolating a virus. It was then characterized both in my laboratory and by Irvin Chen, who had subsequently arrived at UCLA from Temin's laboratory. It turned out to share about 50 percent of its genetic sequences with the earlier HTLV. The new retrovirus was named HTLV-2, and the earlier HTLVs were now definitively called HTLV-1.

Since then HTLV-2 has been isolated again—by Golde and co-workers, by us, by CDC investigators, and by Mark Kaplan and co-workers—from several additional cases of leukemia or lymphomas, again involving the T4-cell. Like the original case, all have been less aggressive than the typical ATL of HTLV-1. Epidemiological studies in London by Robin Weiss and in the United States by my co-worker Marjorie Robert-Guroff, and later by many groups, indicate that HTLV-2 is particularly prevalent among drug addicts in the United States and Europe and that it is increasing. Kaplan and his collaborators believe that different isolates of HTLV-2 vary from one another more than do different isolates of HTLV-1.

Two general processes distinguish a tumor cell: (1) failure of the cell to differentiate properly or to die (immortalization), and (2) behaving as though it were being driven to divide. Retroviruses may be involved in contributing to one or both of these processes. Exactly how they do so varies. In fact, the wide latitude in mechanisms employed by retroviruses that convert normal cells to cancer cells is in my view one of their most amazing features. I have already discussed some of them, such as the rare acquisition of a cell gene (the oncogene) that in time becomes altered in such a way that a virus containing this gene that infects a new target cell will rapidly convert it into a cancer cell. But that mechanism and that kind of retrovirus are, fortunately, extremely rare. The more common animal retroviruses induce cancer by far more gradual mechanisms which are not all yet fully understood. HTLVs would have to be more like these "chronic" leukemia viruses because they took a long time to produce disease and because we knew they did not carry a cell-derived oncogene. What was known about the mechanisms of chronic leukemia induction by such retroviruses of animals?

Some involve the envelope of the virus, but we do not know exactly why or how. Others do their mischief by a process called *insertional mutagenesis,* which involves integration of the DNA provirus near a cellular gene, a *proto-oncogene,* * which when activated can be important to growth of the particular target cell. Somehow the integration of the provirus leads to genetic mutations near the proto-oncogene, which, in turn, causes an abnormal "turning on" of this gene and leads to abnormal growth. But this may be only the first of a series of genetic changes needed for the cancer to develop. One suspects that the great increase in cell growth produced by the virus increases the probability for another genetic change, a genetic accident.

Because the integration of the DNA proviruses into the cell's DNA is not specific but almost random, the more provirus made, the greater the chance for integration near the key cellular gene; and, of course, the amount of provirus being formed is related to the amount of virus replication. I believe this is the reason some investigators were led to think that, if present, virus should be easy to find in humans. They were studying systems in which viremia (virus production that is plentiful enough to show free virus particles in plasma) preceded and was required for leukemia, because the mechanism for the leukemia *required* extensive virus replication.

The human retroviruses would prove different. They employ still another mechanism for leukemogenesis.† The first lead came out of studies of the chromosomal site of integration of the HTLV-1 DNA provirus. (It is possible by molecular techniques to determine the location of integration with precision.)

Whereas with common animal retroviruses, the integration site is roughly the same and must be so in every leukemia (because the requirement for leukemia to develop is the turning on of a specific cell gene by the integrated provirus), Mitsuyaki Yoshida and his colleague M. Seiki made the important observation that the site of integration for HTLV-1 in its target leukemia cells varied from patient to patient.

**Proto-oncogene* is a term often used to describe the oncogene in the DNA of the cell in its normal state, not mutated or modified in any way.

†We call the process of leukemia development *leukemogenesis.*

In any one patient, however, the provirus was integrated in every cell in the same place. This latter result demonstrated that the tumor cells were a product of one single ancestor transformed cell (*monoclonality* of the tumor). Since integration is near random, this also showed that the virus that infected the original ancestor first transformed that cell, rather than the virus coming in the leukemic cells after they had formed, as a "passenger virus." This molecular biological result is a powerful additional argument supporting a direct role for HTLV-1 as the cause of this leukemia.

But one patient might have the provirus in chromosome 5, the next patient in a different chromosome, and so on. The mechanism did not operate by bringing in an oncogene, as expected, but neither was it by insertion near a proto-oncogene and activation of that gene. Therefore the mechanism of leukemogenesis by human retroviruses must be different from what was known then for any animal retrovirus.

My colleagues Flossie Wong-Staal, Beatrice Hahn, and Genoveffa Franchini soon obtained similar results. The conclusions were obvious: HTLVs could not work by insertional mutagenesis. Since their general mechanism (turning on of a cell gene) was similar to animal retroviruses, they must accomplish it by a different pathway. The puzzle began to be solved with the molecular analyses of the HTLV genomes.

In addition to *gag, pol,* and *env*—the three genes that all viruses carry to code for the viral structural proteins—the human retroviruses have extra regulatory genes. In the case of the HTLVs, two were well defined by Yoshida and co-workers in Tokyo, by Wong-Staal and colleagues in my laboratory, and by William Haseltine and co-workers at Harvard.* They are called *tax* and *rex.* It is now clear that in addition to the role of regulating the turning on of the viral structural genes, *tax* in particular is also involved in the activation of some cellular genes leading to T-cell proliferation. Ironically, we had good insights into which cell genes were activated long before we had the results of these molecular analyses. These came from the earliest cell biology studies carried out in our laboratory on these leukemic T-cells.

We already knew that T-cell proliferation was driven by a protein

*Subsequently several other investigators, such as Warner Greene, John Brady, Irvin Chen, Thomas Waldmann, Honjo, Tanaguchi, and George Pavlakis, contributed central information on the precise mechanisms involved.

now called IL-2, our T-cell growth factor, and from our very first studies with HTLV-1–positive leukemic cells we knew that they differed from normal T-cells in that they responded directly to IL-2. No prior activation of these leukemic T-cells with PHA or antigens was usually required, as it was with normal cells. The answer seemed simple enough: normal, resting T-cells do not have receptors for IL-2 until they are activated, whereas these HTLV-1–positive leukemic cells must be abnormal in that they continually express receptors for the growth factor IL-2. Our further studies, and soon far more detailed, elegant, and conclusive studies by Thomas Waldmann and Warner Greene and co-workers, extended and quantified these preliminary results. Soon the molecular and the cell biology experiments converged. *Tax,* a gene then known only for human retroviruses, required by the virus for making the mRNA molecules for HTLV structural proteins, "turns on" not only the other viral genes but also several of the host cell's genes and, in particular, IL-2 and IL-2 receptors.

This leads not only to release of the growth factor but also to its binding by these T-cells, and, of course, this in turn leads to abnormal T-cell growth. I think this chronic on-and-off-again growth of T-cells increases the chance of another genetic change, a mutation, which when combined with the drive to grow due to the presence of IL-2 and its receptor both turned on by *tax,* will give rise to the fully malignant T-cells. But we do not yet know what this other genetic change or changes might be. As with the animal retroviruses that cause leukemia, we only partially understand one step in the neoplastic process induced by the HTLVs. My thinking had been that some unknown others almost surely must be operative. Otherwise how can we explain the fact that by the time the leukemia occurs, expression of HTLV genes (including *tax*) is no longer necessary? At least that is what we and others have observed: when the leukemia developed we did not even detect HTLV proteins, let alone the whole virus, until we properly cultured the cells.

But some totally unexpected and exciting results on this problem came out at our annual LTCB meeting in August 1990. With newer, more sensitive techniques, scientists are finding the *tax* protein in the leukemic cells of the patients. Added to this, Prakash Gill at the University of Southern California reported that he had a few patients with AIDS who also developed T-cell leukemia. They were infected with both the AIDS virus and HTLV-1. He was treating the AIDS by using antiviral therapy (AZT and interferon), a rational approach (as I will discuss in chapter 13), whereas it makes little sense to treat the leukemia with antiviral agents because our thinking is that once a cancer occurs, it is too late to treat the cause. (One cannot treat lung cancer by having the patient quit smoking.)

The surprise: the T-cell leukemia responded and went into remission, and this is a leukemia that has not responded to chemotherapy (except by a very new and special chemical "precision bullet" approach developed by Tom Waldmann and co-workers, which I will soon discuss). These results seem to converge and when one sees convergence in science, it usually soon correlates with a meaningful advance. The results suggest, then, that a key may be the turning on (and keeping on) of the *tax* gene of HTLV-1 so that the *tax* protein is present. The presence of very low level *tax* may not be sufficient to turn on the other genes of HTLV-1 (the "normal" function of *tax*), but may be sufficient to turn on cellular genes that lead to abnormal growth. But what is it that turns on the *tax* gene of HTLV, and selectively so? We do not have the answers.

Other new findings on HTLV reported at this recent meeting also had to do with *tax*. Results from my postdoctoral Fellow Ron Gartenhaus, and earlier and more detailed results and discoveries of John Brady and his co-workers, also at NCI, have shown that *tax* is released by HTLV-1–infected T-cells and can influence and promote metabolic changes, including growth induction of nearby uninfected cells. This phenomenon, release of a viral regulatory protein from cells and its effect on *other* cells, was only just discovered with a related gene, called *tat*, of the AIDS virus (as we will see in chapter 10), and is a new concept in virology, one I expect we will soon learn is of considerable importance in explaining some mechanisms of other viral diseases. Yet we are still left with the inability to explain the long interval between the time of infection and disease development.

In the early period of the HTLV work, Miyoshi reported the peculiar presence of antibodies in sera from some Asian monkeys that strongly reacted with HTLV-1. The implication, of course, was that these monkeys were infected and were chronic carriers of HTLV-1 itself or of a closely related virus. I first heard him discuss the results at a meeting in Buffalo in 1981. Because the immune fluorescence technique he used was known for its false positives and because of the oddity of the result, most of us felt his results were probably wrong, but too interesting to ignore. Since I had just postulated a likely African origin of HTLV-1 because of some of the epidemiological results, we tested sera from monkeys but not just Asian monkeys.

We quickly confirmed Miyoshi's positive Asian monkey serum antibody findings, and along with others, such as Gerhart Hunsmann and his associate in Göttingen, extended them to show that although New World monkeys were negative, many African monkeys also contained such antibodies in their sera. Soon some virus isolates were obtained and analyzed. These results showed that the African viruses were more closely related to the HTLVs than were the Asian isolates. Of particular note, Yoshida showed that viruses from chimps and from African green monkeys were about 95 percent identical to the HTLVs. In other words, the simian viruses (now named STLV-1 and distinguished by abbreviation—for example, STLV-1_{CH} for STLV-1 from chimps) from Africa were almost the same as the human virus.

I think these results support my suggestion of an African origin of the HTLVs, a suggestion initially based on the findings of a high prevalence of infection in sub-Saharan Africans and in some black populations in the Americas, and a very low incidence in other groups in the Americas and in almost all groups from Asia. I suggested that perhaps HTLVs were brought to the southern islands of Japan by sixteenth-century Portuguese adventurers who came to that region with people and animals from central Africa. But later I learned of a pocket of infection among the native population in an area of Apulia in southeast Italy and among some American Indians, and of a low rate of infection in Spain, which could not be readily explained. More disconcerting was the finding of Japanese epidemiologists of a very high rate of infection of the Ainu people in Hokkaido, the major northern island of Japan. The Ainu are not only racially distinct from the bulk of Japanese and geographically removed from other endemic areas, but they had not had the Portuguese contacts.

Obviously, this does not eliminate the African hypothesis, as a few have rashly written, but it complicates it. If it were accurate and if the Ainu results are verified, the origin and spread of the virus must be more complex. To be correct, at least two old pathways of travel of the HTLVs must be invoked—the one I have proposed and a much earlier branching of virus and of people, into which I have no possible insights. As far as I know, there have been no publications on the results of molecular analysis of the HTLV-1 of the Ainu. In view of the Japanese tendency for thoroughness, I wonder whether the Ainu really are infected with HTLV-1 in the incidence suggested or whether the antibody assays have detected something related to but differing from HTLV-1.

By 1982, no serious scientist could doubt the existence of human retroviruses. The arguments in favor of their being a direct cause of a human cancer had grown abundant. First, just finding a retrovirus of this kind (similar to the new type-C retroviruses of animals) was suggestive in and of itself of a cancer agent because of the significant precedent of this exact kind of virus causing animal leukemias. Second, the epidemiology was straightforward. These T4-cell leukemias were rare. The virus was uncommon. The chances of these two joining forces at random, particularly in some populations, would be highly unlikely, but they occurred together regularly.

Third, in one particular T4 malignancy, the typical ATL, an HTLV was almost always identified. Fourth, the viral sequences were integrated in a clonal fashion in the tumor cells, demonstrating infection of the first transformed cell, not later infection as a passenger virus. Fifth, in laboratory culture experiments, infection of normal T-cells by an HTLV converted the cell into one with permanent growth. The favored target was the T4 lymphocyte, the same target found in instances of disease *in vivo.* Sixth, some of the monkey viruses (STLVs) have been shown to be associated with lymphoid malignancies in monkeys.

Owing to their capability of immortalizing infected target T-cells in laboratory culture flasks, human retroviruses were also becoming important tools for the experimental study of the mechanisms of conversion of a normal cell to a transformed one. Although there were already other laboratory systems for this, few if any can also utilize the known causative agent of the tumor in humans. We also knew that the HTLVs were spread by products of twentieth-century behavior—an enormous increase in travel, sexual promiscuity, intravenous drug addiction, and increased medicinal use of blood and blood products—so they were likely to increase in the years ahead.

Thus, we could now perhaps almost leisurely think of more academic issues related to these viruses. Could we unravel the molecular mechanisms of the later stages of cancer development by

HTLV? How did they cause neurological disease? How did they cause, or at least increase the incidence of, some other diseases recently suggested by clinicians and epidemiologists? Was HTLV-1 involved in still other diseases? Would there be more human retroviruses? If so, for what diseases? It was a calm time of reflection before the storm.

III

The Discovery of a Third Human Retrovirus: The AIDS Virus

8

A Single Disease with a Single Cause

In the beginning . . . there was no plague, none at all. The very word was forbidden.

Next came the talk of "pestilence"—that is to say, it was a pestilence, but only in a certain sense; not a true pestilence, but something for which it was difficult to find another name.

Last of all, it became a pestilence without any doubt . . . but now a new idea was attached to it, the idea of poisoning and witchcraft, and this corrupted and confused the sense conveyed by the dreaded word.

I do not think that it is necessary to be deeply versed in the history of words and ideas to see that many of them have followed a route similar to that just described. . . . But in small and great matters alike, it would often be possible to avoid traveling that long and tortuous route, if people would only follow a method which has been recommended to them long enough—the method of observing, listening, comparing and thinking before they begin to talk.

But talking—just talking, by itself—is so much more easy than any of the other activities mentioned, or all of them put together.

—From Alessandro Manzoni's discussion of the plague in Milan in 1620 in his classic nineteenth-century novel *I Promessi Sposi (The Betrothed)*

Many times throughout the history of humankind, epidemic disease has mysteriously appeared, terrorized and decimated large populations, and then disappeared, often just as mysteriously. Medical and literary records document the terrible devastation caused by these frightening and unexplained outbreaks of disease. In most such cases, contemporary medical people described them and their causative agents as new, but we are in no position to know for sure. We do know that at least some of the sudden epidemics subsided, then reappeared at a later time. This was the pattern, for example, of bubonic plague, the most infamous epidemic disease. Although there are rational theories for its appearance/disappearance/reappearance —most based on shifts in the relationship between rats and humans— there is no way to test these theories to the satisfaction of a modern scientist. What *is* clear is that epidemic disease is often precipitated by major societal changes, such as fluctuations in population density, new relationships between human and animal, or (and chiefly) new ways of interacting among individuals or groups of individuals. History gives us a number of examples.

Around 430 B.C., a devastating epidemic in Athens was detailed by the Greek historian Thucydides. He believed the disease to be new. It came and went within one season, presumably, according to the historian William McNeill,* because the small surviving population comprised only those who had developed a sufficient and lasting immune response. There were grave population losses and some have suggested that this epidemic, though short-lived, may have, by its severity, altered the course of Mediterranean history.

Its onset seems to have corresponded with a period of increased population both in Athens and in some of the other Mediterranean port cities, and with more frequent contacts among these cities. Thus, in populations that had been virgin—previously unexposed to the offending microbes—and therefore ripe for serious infections from outside sources, widespread infection could be introduced by these new contacts.

Epidemics of the Mediterranean area were recorded in the second and third centuries A.D. Medical historians now believe these were incidents of measles and smallpox. The first major epidemic to fit the

*In his superb 1976 book *Plagues and People,* McNeill discusses many such epidemics and defines the role of specific societal changes in the initiation or reintroduction of each.

description of bubonic plague extended on and off from A.D. 542 into the eighth century.

Still, there were long intervals even in this period during which Europe and the Mediterranean areas seem to have been free of serious epidemic disease. Indeed, entire centuries appear to have been untouched by significant major pandemics. Beginning in 1346, however, bubonic plague struck Europe a second time. So virulent was this outbreak, called at the time the Black Death, that the generic word *plague,* meaning "stroke" or "wound" and previously used to describe any sudden medical calamity, came to apply exclusively to this epidemic disease. Its impact on European life was immeasurable.

The first recorded *epidemic* of syphilis, in 1495, corresponds with the arrival of the French army in Naples and offers an amusing illustration of how history can be written. Whereas most of us would enjoy having our names associated with a new phenomenon, few enjoy being credited as the originator of a new disease. So, we are told, the French, giving the Italians their due, called syphilis the Neapolitan disease, while the Neapolitans, being modest, called it the French disease. In time, both began calling it the Spanish disease. However eager these Europeans were to honor their close neighbors, some historians argue strongly that the timing of the 1495 syphilis epidemic confirms that it was brought to Europeans by Columbus's men. Whatever arguments support this interpretation, they do not easily explain the disease's wildfire progress in Naples so soon after the return of Columbus.

By the sixteenth century, Europe seemed to have gone through the worst of its plagues. Its contact with most microbes was by then old, extensive, and varied. Unfortunately, for the American Indians, it was not, even allowing for syphilis as the one notable exception. The near-eradication of the microbially inexperienced American Indian is perhaps the most dramatic and best-known example of the destructive capability of allowing sudden social change to introduce disease to a new population.

McNeill quotes a reference source that claims that older Amerindians denied having *any* disease prior to the arrival of Europeans! While I do not think that such a claim should be taken literally, it does illustrate the perceived extent of the changes in the Amerindian culture wrought by the many infections Europeans brought to the Americas: smallpox, diphtheria, measles, mumps, tuberculosis, malaria, and yellow fever. Their trial histories record the terrifying impact of these devastating diseases on their lives and cultures.

Major epidemics of malaria and yellow fever hit the New World in the seventeenth and eighteenth centuries. Europeans in America suffered extensively, leading some historians to speculate that both these epidemics had an American origin. More careful review of the existing records, however, strongly indicates that they were new both to Europe and to the Americas, reaching the latter from Africa via the slave trade. (Among African primates, for example, unlike their American and Asian cousins, the malaria parasite and its mosquito vectors are highly evolved and specialized, suggesting a long history of adaptation and an African origin for malaria.)

In the mid-nineteenth century, a pandemic of cholera raced across Europe, attributed to British military excursions into India and a marked increase in European shipping contacts with Asia. As the prevailing balance of the cholera microbe and the native population was upset, the microbe spread widely. Growing industrialization, with its accompanying increase in city dwelling, offered an opportunity for this epidemic to emerge, particularly in England but also elsewhere in Europe.

The first serious pandemic of the twentieth century was the 1918–19 attack of influenza. Although of unclear origin, it broke out in many new locations at the end of World War I, strongly suggesting that it was the result of the coming together of troops from several continents.

Epidemics that seem to have begun suddenly and to have disappeared just as suddenly can, in fact, be either truly new or a recurrence of an old disease gone dormant in the population over many centuries. Tularemia, an often fatal microbic disease transmitted by rabbits, was frequently reported in the early part of this century, but rarely before then. We don't know why. Polio was first identified as a disease in 1840, but records of medical examinations suggest it existed earlier. Why it was never identified earlier as a distinct disease is a mystery; this failure may well have contributed to its reaching epidemic proportions in the nineteenth and twentieth centuries.

Another more recent example is Lyme disease. Described for the first time in the 1970s, this infectious, tick-borne disease is caused by a kind of bacterium known as a spirochete (similar in some respects to the microbe of syphilis; both have a corkscrew shape, giving them their formal generic names). The origin of the disease is unknown.

In 1976, Legionnaire's disease first attracted wide public attention with a sudden outbreak in Philadelphia at a meeting of members of the American Legion. With stellar work by David Fraser of the Centers for Disease Control, the microbe and its method of transmission were both nailed down. Airborne bacteria, named *Legionella,* were growing in the air-conditioning cooling towers of the hotel where the Legionnaires were staying; the bacteria spread in aerosol form through the air ducts. Here was a microbe that was not necessarily new itself but whose transmission was facilitated by modern technology.

Common sense tells us that all diseases had to have been truly new at some point in history, but many epidemics that appear to be new diseases may simply be newly recognized in the population describing them (as were, for example, yellow fever and malaria among Europeans and colonial Americans), or may have reached epidemic proportions for the first time (the attack of syphilis in 1495 probably falls in this category). It is a rare disease that is truly new when first documented.

Probably starting in the 1960s, perhaps even earlier but not much earlier, a previously unreported epidemic disease that was difficult to transmit and that had a long period of apparent latency was silently but relentlessly establishing itself. Its exact place of origin may not be fully certain, but the time of its appearance seems more settled. With extremely few exceptions, pre-1970 sera saved in laboratories all over the world were found to be negative for the telltale sign of the agent of this new infection.

Within a decade this new disease reached epidemic proportions in parts of equatorial Africa, and the first cases appeared roughly simultaneously in the United States, Europe, Haiti, and perhaps a few other areas of the world. The spread of the disease relied upon precipitating social factors unique to the twentieth century: an enormous increase in international travel; an increase in individual sexual activity, especially with multiple partners; the accelerating movement of blood and blood products for medical purposes not only from person to person but from nation to nation; and the social insanity of intravenous drug abuse. These factors made possible the global spread of a microbe that is distinctively difficult to transmit. But another factor also contributed to that phenomenon: the very nature of the causative microbe. This virus inserts itself into its target cells in a way that makes the infection lifelong. Since its effects are slow, the individual remains infected for many years. Thus he or she is capable of infecting others not just for days or weeks, as with most other viruses, but for years.

Initially, the disease was difficult to identify as a new epidemic because it took many forms. By attacking the white blood cells and the body's immune system, it made its victims vulnerable to a variety of opportunistic infections, many of them fatal. In addition to its distinctive period of long latency and the variety of its presenting symptoms, another trait uniquely characterized this newly emerging epidemic disease: its agent of infection was unforgiving. Few, possibly none, of those who contracted it survived with a natural immunity.

In the United States, the first individuals to show signs of infection were virtually all homosexuals, living in either New York or San Francisco. The most frequently observed symptoms—weakness, chills, enlarged lymph glands, and, surprisingly often, purple skin blotches characteristic of a type of slow-growing cancer previously found only in older men of Mediterranean ancestry—yielded one additional clue to doctors: a precipitous drop in the individual's white-blood-cell count, particularly the CD4 lymphocyte count. Within a few years, the disease we now know as AIDS would begin to devastate the homosexual community. Soon thereafter, we realized that anyone who required a blood transfusion was at risk of getting it as well.

A truly new microbial disease of humans can come about in one of only three ways: the entry into humans of a microbe that has previously established itself in another animal; a genetic change in a microbe already in the human population that makes the microbe more capable of survival, more infectious, or more pathogenic (disease-causing); or a change in humans that makes us more susceptible to a previously innocuous microbe. A disease not truly new to humankind can still present itself as new to a given population or even to most populations if the microbe spreads to those populations from one in which it has always been confined. Of these factors, one stage or another of the spread of AIDS worldwide was possibly, and perhaps probably, due to the microbe's transmission from animal to human and from a remote section of the world to the developed world.

I read in the newspapers about this new disease and listened quietly when discussions developed at NIH about its possible cause. But at the time I was still involved in our HTLV work and some other aspects of cancer research; I was not yet listening with a professional ear. It would be close to a year after the first reports in the medical literature that I would take a closer look at this disease, whose agent of infection by then had started to show close parallels to what we knew of the HTLVs.

Later, it would be suggested that the research establishment, myself included, had been slow to show interest in the disease because it seemed to affect primarily the homosexual population. Such thinking shows a failure to understand how scientists choose what to work on. Scientists are drawn to problems that seem capable of solution given the present state of knowledge, specifically those problems where their own abilities and training make a personal contribution possible and address a societal need. I was led into the field of retroviruses very much because of an interest in Peyton Rous's work on a sarcoma that afflicted only chickens, William Jarrett's work showing that a similar virus caused leukemia in cats (rather promiscuous ones at that), and Howard Temin and David Baltimore's discoveries that a special enzyme synthesizes DNA. To say that I failed to be interested in a disease that struck down fellow human beings because the sexual practices of some of its victims may have been different from mine is egregiously absurd.

The first medical account I heard of the disease that came to be called AIDS came in the summer of 1981, at UCLA. Michael Gottlieb and his co-workers reported a cluster of pneumonia cases in young male homosexuals in Los Angeles caused by a protozoan parasite known as *Pneumocystis carinii* (PC), a common but normally not pathogenic microbe. Previously, PC was found to have caused pneumonia only rarely, and then in severely immune-depressed people. It had been observed, for example, in those whose immune systems had been intentionally weakened by medical therapy as part of the preparation for organ transplantation.

Almost at the same time, Alvin Friedman-Kien and co-workers at New York University Medical School found a cluster of previously quite rare Kaposi's sarcoma (KS) cases among homosexuals, and a similar cluster of disease in gay men in New York was reported by Fred Siegel and co-workers at Mt. Sinai School of Medicine in New York. The Kaposi's sarcoma disease showed up as purple, cancer-like lesions of the skin. Still other clinicians began to report lymph gland enlargement in young male homosexuals, as well as an odd increase in an unusual B-cell lymphoma (a cancer of the lymph glands of B-cell type, which I have touched on in earlier chapters but not in its association with AIDS).

These extraordinary findings did not go unnoticed by the Centers for Disease Control in Atlanta, whose role it is to monitor disease patterns. James Curran at the CDC and a few other epidemiologists believed they might be looking at the first signs of a newly emerging and potentially epidemic disease. I soon became aware that clinicians were reporting another unusual finding, one that would prove particularly intriguing: young male patients with a reduced number of circulating T-cells and in particular a reduction chiefly in the T-cells known as T4 lymphocytes or the $CD4^+$ T-cells.*

In late 1981 Curran came over to NIH to talk about these new epidemiological findings. His purpose was not simply informational. If all this confusing data were in fact defining a new disease, and if the disease were as serious as he suspected, it would be necessary to get NIH talent working on it. But who could make the most useful contribution: researchers in infectious disease, virologists, oncologists, immunologists?

Curran provided us with a description of what he considered the range of sexual practices between males—not for voyeuristic reasons, but because of the possibility that hidden within their patterns of sexual activity might be a clue to understanding the disease. The early epidemiological findings pointed to some connection between male homosexuality and the disease, though no one yet understood just how any of the sexual practices of gays—either the type of practices or the frequency of contact—could contribute to the cause or spread of the disease, if indeed they contributed to either. Though no one who heard Curran speak or followed reports in the press could fail to be moved by the horror of what this new disease was doing to young male homosexuals, nothing Curran said that day caused me to believe that this was an area of research our lab should move into.

But Curran returned in early 1982 with far more—and far more compelling—epidemiological information. If my memory is correct, it was at this time that I learned there was reason to believe that the disease was showing up in people who had had blood transfusions,

*T4-cells are now usually called $CD4^+$ T-cells by medical scientists. Thus, the terms T4 and CD4 can be used interchangeably. CD4 is a specific protein molecule on the surface of some mature T-cells. It has an important function: it helps "bridge" the T4-cell to some other cell types (like macrophages) by its loosely binding to a specific protein on the surface of those cells. This helps the two cells exchange signals during the immune response. The reader will remember that the T4-cell is also the very subtype of T-cell chiefly targeted by the HTLVs.

implying a microbial agent. As well, there was by now clear evidence that it compromised T-cell function, an area of study closely related to our work on HTLV.

Many of the theories about how the disease was spread—ranging from the reasonable but unimaginative, to the questionable but creative, to the outright bizarre—did not presume an infectious origin. Yet Curran seemed nearly convinced that an infectious agent was involved, probably a new one. He had no data suggesting what the agent might be, or be like, and never speculated in this direction, but here is where he thought we ought to start looking, and start fast.

This second talk did shake me up and led me to think very carefully about AIDS. Frankly, it is my belief that Curran's efforts and those of a few others through this period, including Dean Mann at NCI and Bob Redfield at Walter Reed Hospital, were central to the first appreciation of the AIDS problem by much of the medical world. I shudder to think what might have occurred had they and others not forced what turned out to be an eleventh-hour warning on a research community that sometimes sees itself as too tied up with previous research commitments to undertake new ones.

I left Curran's conference stimulated and concerned. Of greatest interest to me was that this new disease, likely infectious and threatening to become an epidemic, appeared to involve T4-cells, those white blood cells of the lymphocyte lineage. The precise method of transmission in gay men had not been established, but the vehicle of transmission in all other cases seemed to be blood. Later we would learn that it was also transmitted from mother to fetus (or mother to child, if the infection occurred during the birth process). Shortly thereafter I happened to speak with my collaborator Max Essex, who reminded me of several important studies that had shown a correlation between some animal retroviruses and the suppression of the immune system. William Jarrett, for example, had observed more than a decade earlier that the feline leukemia retrovirus more often caused an AIDS-like immune deficiency in cats than it did leukemia. Max had confirmed and recently extended these results, and we wondered aloud whether the disease in cats might be caused by a small mutation in the genome of the feline leukemia virus.*

Essex had also told me about his recent study in Japan showing that

*In 1987 we learned from the work of James Mullins that this is almost exactly the case. Certain changes in the envelope of the cat leukemia virus convert it into a virus that can cause an AIDS-like disease.

there were more HTLV-1–positive people in infectious disease wards than in other medical wards. Conversely, research by Takatsuki found more opportunistic infections in leukemic patients infected with HTLV-1 than one would expect. And several labs—including mine (with my co-workers Marv Reitz and Mika Popovic); David Folkman and Tony Fauci at NIH; Bo DuPont at Sloan-Kettering Institute in New York; and Nicole Suciu-Foca at Columbia—had published evidence that HTLV-1 could be immunosuppressive *in vitro:* that is, it could harm the function of T-cells in laboratory cultures.

I started reviewing all we knew about HTLV in the new light of what we were learning about this new disease. The HTLVs were transmitted by blood and sex, as well as gestation, the birth process, and mother's milk. HTLVs infected T4 lymphocytes and could reduce the immune function of T4-cells. Was AIDS caused by a retrovirus? Was this retrovirus in fact a variant of HTLV-1 (or 2)? Was it perhaps a new retrovirus—one related to the HTLVs?

Intellectually, I began to play out one scenario. What if AIDS were due to a mutation of an HTLV, probably occurring in Africa, which had spread to Haiti, then to the United States? The feline leukemia virus (FELV) mutation idea, if proved correct, should make us think about a similar possibility with the HTLVs—if, in fact, the causative agent did turn out to be a retrovirus. Perhaps the differences between the way an HTLV caused T-cell leukemia and the way a related retrovirus caused AIDS might be due to a small change in those extra regulatory genes (called now *tax* and *rex*) that Yoshida in Japan, the Harvard group, Wong-Staal and others in our group, and later other groups had found in the HTLVs but that were not known in any animal viruses. Or perhaps a change had occurred in the envelope by some mutation or other means of genetic alteration. Differences in the envelopes of various strains of the same animal retrovirus can lead that virus to cause different diseases. A lot of ifs, but that's how we decided where to start to look for confirmation of an idea.

But at first I hesitated to plunge in. Looking back, I see clearly now that part of what was holding me back was the memory of how much trouble we had had getting the scientific community to accept the existence of even one human retrovirus. Now we had two—HTLV-1 and HTLV-2—and there was still plenty of work to be done on them. It was tempting simply to continue with that work now that it was going well. I decided to sound out an old friend, a bit of a tough egg who had more experience than I in the scientific life.

Sometime in early 1982 I met my close friend, the late GianPiero DiMayorca, for dinner in Bethesda. Born and raised in Milan, DiMayorca was a highly intelligent, handsome, brash, but very—perhaps too—cynical virologist. In the early 1980s he became chairman of microbiology at the New Jersey School of Medicine and Dentistry in Newark.

By all accounts, he argued, AIDS was a dreadful disease; if it followed the pattern of past epidemic diseases, it might be years, maybe centuries, before a cure for it would be found. If it turned out to be infectious, which seemed more certain every day, how infectious would it be, how ingenious its agent of infection in getting around the defenses that had evolved—both physiologically and culturally—to help us deal with other dangerous microbes? And what special risks would there be in studying the illness and in handling specimens?

Despite the extreme caution that is part of the environment of any good lab, my friend pointed out, other microbes had in the recent past infected laboratory workers attempting to isolate them. He mentioned Marburg virus and other instances where an unknown agent had been brought into a laboratory and had harmed or killed virologists or their technical staff.*

There was a further frightening aspect to this new disease. It appeared, so far, that no one spontaneously recovered from it. It did not run its course over a short period of time, reach a critical stage, and then either kill the patient or leave him or her recovered with some natural immunity to new infection. It seemed to be killing everyone who came down with it. If I brought it into my lab, I risked not only my own life but the lives of my co-workers. I might be spending the next decade or so watching one former colleague after another come down with those first symptoms that would mark him or her as having contracted a surely fatal disease.

Having put these ghoulish thoughts into my head, DiMayorca then turned toward the professional aspect of doing research on AIDS. Patients with AIDS, he reminded me, were infected with so many different microbes (either because of lifestyle or because of the suppression of their immune systems) that establishing any one agent as the sole cause could prove frustrating, perhaps impossible. In short,

*Marburg virus belongs to a new class of RNA viruses. It was discovered in 1967 after an outbreak in Europe in which thirty-one people were infected and seven died. It began when virus from African green monkeys infected laboratory scientists who were working with the monkeys' kidneys.

the whole dangerous project showed slim prospects for professional success when measured against its downside.

"If you are right about the retrovirus idea," DiMayorca said to me, "let someone else do the dangerous work. You can sit back and watch, knowing you and your staff originated the notion and already found the first human retroviruses. If you are wrong, you have much to lose; if correct, little to gain. Let them turn to you and your group as experienced advisers. Your people have earned this. Also, you are a cancer researcher in the National Cancer Institute. This new epidemic is not a priority for cancer research, even though Kaposi's sarcoma appears to be a part of it. At least it's not a priority yet."

Neither my friend nor I considered at the time the possibility that did come to pass—that I would fully embrace work on AIDS, that I would turn out to be, in the main, right, and that my reputation would still suffer. Ours would become the most influential lab in the AIDS field, putting itself out front on the retrovirus theory; we would turn out to be dead right about a retrovirus being the culprit; we would contribute substantially to the identification and culturing of the particular retrovirus that causes AIDS; we would be the first to grow the AIDS virus in sufficient quantities to begin serious work with it; we would be the first to develop a workable blood-screening test for AIDS; we would produce much of the information on the basic makeup of the virus; we would provide most of the results that showed the new virus to be the cause of AIDS; we would begin to understand the relationship between AIDS infection and the more aggressive form of Kaposi's sarcoma; we would discover a new herpes virus as a possible co-factor. And yet despite all this hard work, by myself and my colleagues, I would find my reputation attacked in the press coverage of a patent suit between the United States and French governments—a lawsuit, I might add, to which I was not a party, over a patent I had not requested and from which I never expected any financial gain.

Ultimately, the suggestion would be made (and at times mindlessly repeated, until responsible people who should have known better would demand an investigation of me because of it) that the Pasteur Institute in Paris had done all the work and that the Laboratory of Tumor Cell Biology had either misappropriated or somehow stolen their AIDS virus.

Of course, in 1982, I had no crystal ball and had to go with my feelings. DiMayorca's arguments were not without basis, but even before I met with him I had already decided to devote at least some lab time to AIDS research. Curran had caught my attention. Citing the earliest epidemiologic findings of the CDC, he put forth the idea that this seemingly random collection of symptoms, which was still being characterized as a syndrome, would turn out to be a single disease with a single cause. In making this prediction, he and a few others would be correct and way ahead of most others then thinking about AIDS. For myself, of greatest importance was that I suspected he was right. Within a short period of time, I would be prepared to go a step farther and suggest that the agent most likely responsible for this disease was a virus, and more specifically a retrovirus.* At our next staff meeting, I suggested that a few in the lab get tissue and begin cell culturing while others might want to do molecular analyses of DNA obtained directly from tissue of AIDS patients—using probes from the HTLVs to see whether we could find sequences related to an HTLV.† No one demurred.

And so we began. In May 1982, Ersell Richardson, a technician in the laboratory, started the first cell culturing. She would later be joined by Betsy Read-Connole. Our strategy was straightforward. First we would get blood samples from three categories of people—confirmed AIDS patients, asymptomatic people in groups the CDC had found to be at high risk, and healthy people not in high-risk groups (controls). We would culture T-cells from all three groups with Interleukin-2 (IL-2) to get them growing. Next we would look in all the cells for the presence of a retrovirus by determining RT activity. If RT activity were present, we would have reason to believe we had a retrovirus and would then try to determine the relatedness of the virus to HTLV-1 and HTLV-2. If the results were inconclusive, we'd then try molecular probing, that is, we'd see if sequences found

*If this idea proved wrong and no retrovirus caused AIDS, there might still be useful information coming from such studies and possibly even some discoveries. For example, AIDS patients are often infected with a wide variety of microbes. We might discover a new virus (including, possibly, a new retrovirus) previously missed because only in immune-suppressed people would this retrovirus replicate sufficiently for us to detect it. The new virus might play no causative role in AIDS but might cause some side disease and/or make AIDS progress more rapidly as a co-factor. Second, if HTLV-1 or HTLV-2 were found (and not a variant of these or a new retrovirus), it would be useful to know how often. Were these already known viruses, which cause disease on their own, spreading in some populations (a not unimportant question even if the HTLVs were not causing AIDS)?

†Molecular probes, in this case, are segments of the DNA proviral genes of the HTLVs tested with DNA from AIDS patients' blood cells and from Kaposi's sarcoma tissue for the interaction known as molecular hybridization. A positive result means that in some cells a virus is present.

in the DNA of the virus-infected cells showed HTLV patterns. If they did, and if they appeared frequently enough in the DNA of AIDS patients to be of interest, we could then molecularly analyze the retrovirus to determine whether it looked like a variant of HTLV-1 or HTLV-2, as we anticipated—or, perhaps I should say, hoped—it would.

As I look back on this period, I believe that Betsy and Ersell were among the first researchers to culture cells from AIDS patients regularly. They plunged ahead at a time when Americans were terrified of AIDS (and no one had yet given them any reason not to be), when a number of medical research institutions in the United States didn't even allow their researchers to work on specimens of AIDS tissue, and when, even at NIH, concern was expressed about bringing AIDS specimens into the laboratory. Ersell would later comment to Mark Caldwell, a Fordham professor who was writing an article on the lab: "Betsy and I were working on cells that other people wouldn't even handle. . . . We enjoyed it, and anyway, if you're the type of person who worries about breathing in virus every time you walk through the door, you can't do research."

As a consequence of our early start, we very much defined the conditions for working with the AIDS virus. Intuitively, one might think, the more precautions the better, suggesting that we should work with upper-level containment facilities, what we call here P-3 or P-4. But I hesitated to do so. Under these stringent conditions, a lot of time is spent changing work clothes, going in and out of showers, working with your arms hooded as you reach into incubator-type equipment. This is less than convenient, and in any case it is only in recent years that we obtained facilities that approach P-3 levels.

My own logic, based on my experience with the HTLVs and past clinical-epidemiological studies, indicated to me that the AIDS virus was not going to be a virus we would get through the air. But we could get it from cuts or from broken skin. So the one rule I made was no glassware or needles in the laboratory. It was also decided that our people would be tested at least once a year. Fortunately, nearly a decade later, no one in our lab has ever tested positive. Sadly, that is not true of all laboratories working with this virus.

Years later the most prominent memory Betsy Read-Connole would hold of those years was not the possible danger she faced personally but the fact that she never knew when new samples would arrive or in what condition. She had to be in the lab virtually day and night—and, whenever out of the lab, immediately available—because when fresh cells did arrive, they had to be started in culture as early as possible. Her sharpest memories were of her frustration when "we'd get a phone call and someone would say, 'Oh, we have a package for you. It's sitting in such and such an airport, and it's been there for three or four days.' "

In fact, such delays in getting material to the lab distorted our early findings. We rarely received fresh cells, a problem whose seriousness would not become apparent to us until we realized just how cytopathic* the AIDS virus is to its target cell: *in vitro* it replicated quickly upon infecting a host cell and brought on cell death not long after. For quite some time we would test for the presence of reverse transcriptase activity and usually find nothing, and all these false negatives threw off our numbers.

The only other person from the lab working directly on AIDS research at the very beginning was Ed Gelman, a clinical hematologist working with Flossie Wong-Staal in our molecular genetics group. Though he never cultured cells to attempt virus isolation, his experiments were actually the first in AIDS research to be carried out by our group; they date to the end of 1981 or the earliest part of 1982. He began using molecular probes of HTLV-1 and HTLV-2. When positive results were obtained, the DNA was treated with special enzymes called *restriction enzymes.*

Scientists had earlier discovered that these enzymes act much like a pair of molecular scissors, that is, they cut DNA. These enzymes always make their cuts after reading one particular, very short nucleic acid sequence, no matter where or how many times in the genome that sequence is found. By comparing the lengths of the cut pieces, scientists have an idea of how similar, or dissimilar, two strands of DNA taken from different sources are.†

Gelman began looking for nucleic acid sequences in the DNA of AIDS patients or patients with Kaposi's sarcoma that were closely related to the nucleic acid sequences of HTLV-1. Gelman's work also marked the beginning of our lab's long-term commitment to finding

***Cyto* = "cell"; *pathic* = "disease." *Cytopathic* here refers to the virus harming or killing the cell.

†This technique gives a researcher more confidence, providing more information than did older techniques, which were limited simply to the positive hybridization (or matching) of the probe with the test DNA.

out the relationship between AIDS and Kaposi's sarcoma, work that continues to this day.

As Betsy and Ersell cultured their cells, two senior members of our lab began their work. Mika Popovic, a familiar figure in this story, had a particular expertise in cell culturing. As I noted earlier, he had joined the lab in 1980 to work on HTLV-1 and HTLV-2. It was Popovic who had successfully used HTLV-1 to immortalize human umbilical-cord blood cells and made this a routine procedure.*

Now it would fall to Mika to find a way to get the putative AIDS virus to replicate under laboratory conditions—that is, in cells that were themselves growing and were sufficiently hospitable to the virus to allow for repeated viral replication. Prem Sarin, a biochemist as well as my deputy lab chief (responsible for all its financial and administrative aspects, leaving me free to concentrate on science), would help in this end of the work during its earliest phase.

The long years of HTLV research had taught us a great deal about how to handle HTLV-infected cells—how to stimulate them and how to culture them (using Interleukin-2 or by co-culturing them with other cells). We had also learned that the optimum time for expression of virus particles (and hence the best time to test for reverse transcriptase activity) was several weeks after initiating laboratory culture of the cells. In the culturing of the infected blood cells, some time is needed both for the turning on of viral gene expression and for the release of sufficient numbers of virus particles to be detected.

If our initial ideas were right, we were well positioned to make major advances quickly. We had the technological expertise—indeed, we had pioneered some of these processes ourselves—and we had the hands-on experience of working with human retroviruses. The only shadow on all this good news was that if the agent causing AIDS turned out not to be a variant of an HTLV—that is, if it turned out to have major differences—no one knew better than we the long hours of cell culturing that would lie ahead.

First we had to establish that the biologically active agent in AIDS cells was indeed a virus, and then whether it was a retrovirus of any

*We first learned about it from Miyoshi in Japan. Mika had then received considerable and generous help from Dean Mann, an immunologist at NCI, and from Jun Minowada, then at Illinois and now in Japan. He and all my group would receive generous help from them again at an important juncture of our AIDS work, as I will soon discuss.

kind. So Sarin began testing the earliest cultured samples for reverse transcriptase activity by the spring of 1982. If they did not test positive, ever, under any conditions, for RT activity, then no matter how much we believed AIDS was likely to be caused by a retrovirus, we could not proceed further on this premise.

Our earliest RT assay results were ambiguous. Occasionally, we saw some low-level activity, but most samples showed no evidence of RT activity at all and none showed exceptionally high activity. We knew from our earlier experiences with the HTLVs that we were not likely to see much virus production—that is, evidence of free-standing particles—so we were not yet prepared to give up on the retrovirus theory. Nonetheless, what we saw was not yet very encouraging.

There was another, more troubling development: we were having difficulty keeping the cells going in culture. With each sample, we had only a short time to work. Most of the time we had no significant findings to report and simply froze a part of these cultures for later use.

Not surprisingly, over the course of the next twelve months, we struggled, and we learned, as we tried to figure out what was going on and why. Later we would realize how many near-misses we had had. We were not expecting the AIDS virus to be capable of immortalizing cells, as the HTLVs had shown themselves capable of on occasion. Rather, we believed the AIDS virus to be growth-inhibiting. But something more was happening here. What we did not recognize in the earliest stages of this work was that the AIDS virus is not simply growth-inhibiting but *lytic*—that is, it kills cells. This very much distorted our earliest RT findings because the peak period of viral production would usually occur well before we began testing, and cells sent to us from patients at a distance were possibly dying or damaged from the effects of the virus by the time we received them.

All our early cultures, whether or not they tested positive for reverse transcriptase activity, were also tested by immunofluorescent assays for HTLV antigens. Here again our preliminary results were inconclusive. A few cultures tested positive for HTLV-1; most did not. (Those that did show evidence of a retrovirus—by being RT-positive—but were negative for HTLV-specific antigens were, we would later learn, the AIDS virus. Those positive for an HTLV were, we also later learned, infected with both an HTLV and the AIDS virus.) We decided to send out for electron microscopy those samples registering positive for both RT activity and HTLV antigens. Also, because we were working on the assumption that we were looking for a variant of HTLV-1 or HTLV-2, these positive samples were among those sent over to Ed Gelman for molecular probing.

Our AIDS work would later on undergo a degree of scrutiny in one or two elements of the press that I think is unparalleled in the annals of medical scientific history. One of the most absurd statements was that we had ignored the first "sightings" of an AIDS virus in the cells that proved RT-positive, HTLV-negative, because I had too strong an emotional investment in having the AIDS virus turn out to be an HTLV variant. It would even be suggested that much time—and many lives—would have been saved had we immediately moved to work on the other virus(es). This ridiculous view fails to take into account that such samples could not be grown well during the early period. Therefore we did not discuss our work with the samples at this time; this cannot be taken to mean that we ignored them.

How strongly did I think that the cause of AIDS would turn out to be an HTLV variant? Strongly, perhaps very strongly. Nor was I the only serious scientist who thought so. In June 1982 I received a call from Arthur Levine, now director of the Child Health Institute, to let me know that he thought the cause of AIDS might well turn out to be a retrovirus, indeed an HTLV-variant retrovirus. Levine had had years of work in molecular virology and was not a man to jump on any bandwagon. If anything, his intellectual style was to approach new theories cautiously.

To put things in even broader perspective, it is fair to say that every serious scientist actively working on AIDS research during this period, and much later, who felt the notion of a retroviral cause of AIDS was a good hunch also felt the need to look at AIDS in terms of the only other known human retroviruses—the HTLVs.* It would not

*Indeed, long after we knew exactly what the correlation was and how it came about, other scientists were still reporting a variety of curious correlations between AIDS and the HTLVs. I will mention just three. Jean-Claude Chermann, the Pasteur Institute former co-worker of Luc Montagnier, told me that Jay Levy, a research scientist in California, had explained to him in March 1984—nearly two years after the period I am discussing now—that the *majority* of samples he tested from symptomatic AIDS patients showed HTLV-positive responses! A month later, Don Francis, then at the Centers for Disease Control, reported in the English journal *Lancet* a correlation between HTLV-1 and blood-transfusion–induced AIDS based on virus isolation. Bernie Poiesz, my former postdoctoral, working in Syracuse, presented in February 1984 evidence by a then-new and sensitive technique for molecular probing. He found an extraordinary correlation between HTLV-1 and AIDS. The report by Francis, we now know, is of

have made sense to work otherwise. The answers to certain questions must be locked up behind you before you can comfortably go on to the next possibility. Those who do not understand this "rule-out" concept are often confused by what is being attempted by the researcher or clinician (doctors, too, use the concept in diagnosis).

But most important, we proceeded as we did because we had no alternative. At that point we could get neither enough virus nor healthy-enough infected cells to work with. All we could do was use our experiences with the HTLVs as a reference point for dealing with the differences between the known patterns of the HTLVs and what we were seeing. We never expected the AIDS virus to be an immortalizing HTLV; what we were looking for was a variant. In examining how the AIDS virus differed from the HTLVs we would be defining how to proceed. Why, for example, did the addition of IL-2 to AIDS-infected cells not induce cell growth and evidence of viral replication? Why were we unable to co-culture AIDS cells and umbilical-cord blood cells and see viral replication? Why couldn't we find a way to get enough primary culture or virus to try to characterize it and develop reagents? At that point we could not even get low-level viral detection with any frequency. Our problem was not that we were insensitive to the information we were getting about there being differences between the HTLVs and the AIDS virus; it was that we were not yet equipped to deal with these differences and no one else was solving the problems for us.

We did not destroy those RT-positive, HTLV-negative cells. We observed them for as long as we could. Most died before we could learn anything from them. Others were put in our freezer to restudy if and when we learned more. But some survived long enough in culture to reveal something very interesting. Whatever it was that had induced these cells to grow poorly, in several instances we saw evidence that it also caused cells to fuse into what are called *syncytia*—giant cells with many nuclei each. This told us there were biologically active virus particles present that could be damaging these target cells in a way that was very different from the HTLVs.

As far as I know, the only other lab group in the United States

---

occasional (about 10 percent) AIDS cases infected with both the AIDS retrovirus and an HTLV. The report of Poiesz was most likely from faulty methods. I do not know the explanation of Levy's results because he never reported them.

thinking seriously in 1982 and 1983 about a retrovirus as the cause of AIDS was Max Essex's at the Harvard School of Public Health. Toward late summer of 1982, I heard from the French clinical immunologist Jacques Leibowitch, who was with the Université René Descartes, in Garches, France. He had read a medical news report in August 1982 describing my ideas. These intrigued and stimulated him. He asked me whether I really believed that a retrovirus might be the cause of AIDS. I told him my thoughts at that time. Later, I learned from him that he, in turn, had conveyed some of my thinking to other French scientists and to the director of Pasteur Diagnostics, a company with close affiliation to the Pasteur Institute in Paris. He also told me that the Pasteur Institute scientists Jean-Claude Chermann and Luc Montagnier and their co-workers, partly as a consequence of these discussions, had started looking for a retrovirus as the cause of AIDS.*

Chermann was a retrovirologist, having worked with mouse retroviruses for many years. In fact, some of his training was with Peter Fischinger at our institute. Montagnier had little previous experience with retroviruses and, to my knowledge, none with human retroviruses. He had worked mostly with DNA viruses, but in the late 1970s and early 1980s, his work focused on the use of interferon in treating some diseases.

By coincidence, a key technician of Montagnier, who later obtained her Ph.D. degree, had just finished working with us and returned to Montagnier's group. One of their key cell-culturing people was Françoise Barré-Sinoussi. Françoise is both a personal friend and a valued colleague. Then (and for many previous years) she worked closely with Chermann. She too had just spent time in my laboratory—a few months—doing cell biology with B lymphocytes. Chermann and Barré-Sinoussi were open, warm people; our relations were those of friends.

This group at the Pasteur Institute would begin their work in December 1982, moving in the same direction as our lab with, not surprisingly, precisely the same approach. One month later, they would receive a specimen of lymph node tissue from Frédéric B, who is usually referred to as BRU, an abbreviation of his name. (In time,

*I should immediately add that Leibowitch's own independent ideas were also consistent with exactly this kind of virus.

his cells and serum, and even later virus from his cells would often be called BRU.) The French were able to keep the culture going long enough to show reverse transcriptase activity and later to transfer the virus to healthy T-cells. But the culture kept dying out on them and they were concerned about losing it altogether.

In January 1983, Chermann called to tell me about their positive reverse transcriptase in this one sample and to ask for my advice about how to keep their culture going. I suggested that he add human T4 blood lymphocytes obtained from the umbilical-cord blood of newborns as target cells, an approach used earlier for HTLV-1 by Miyoshi and by our laboratory. I also told him our protocol for HTLV-1 and HTLV-2. Applying this information he succeeded in saving their virus (which would later be called LAV/BRU, for lymphadenopathy virus of patient BRU), by obtaining and adding new umbilical-cord blood cells every few days.*

By the end of 1982, there were many theories being touted by a wide variety of people about what caused AIDS. Max Essex and I talked extensively during this period about our own developing thoughts, but except for very brief comments at meetings, one of which had been published in summary form in August 1982, neither of us detailed these ideas publicly until the Banbury Conference on AIDS in February 1983. To my knowledge, this, in fact, was the first meeting of a substantial number of basic scientists gathered specifically to discuss the cause of AIDS.

Their ideas ranged widely. Many still invoked a noninfectious cause. One leading theory was that the primary cause was sperm being released into the blood. The victim made antibodies to the foreign sperm, some of which cross-reacted with $CD4^+$ T-cells and harmed or killed these cells (an autoimmune mechanism). I noted aloud that heterosexual women had been exposed to sperm without triggering such a reaction, and that recipients of blood transfusions who had contracted AIDS had not, as far as I knew, also received sperm. It was, of course, possible and not at all unreasonable that sperm, by entering into the bloodstream through anal intercourse and trauma, played some role in triggering, or hurrying along, other processes. Perhaps

*But Montagnier has told me and other people that *he* did not need this help because he had our published work and knew it very well.

this is the way the originators of the sperm idea thought of it. But it could not be the *sine qua non,* the primary cause.

Another popular idea was "antigen overload." A seductive hypothesis no less off the mark than the sperm theory, antigen overload purports that there is *no* specific cause of AIDS. Rather, a host of infectious agents overload the immune system. Thus a change in lifestyle would eradicate the disease. The theory offers AIDS patients hope and relieves them of responsibility for acts they engage in that might later be found to be destructive. It is a notion that has been resurrected almost seasonally in a myriad of ever-embellished forms. Obviously, this concept cannot explain the sudden onset of this epidemic, nor account for its global spread. And there remained the blood-transfusion cases and the newborns. Were all of them antigen-overloaded? To my mind, it was not only an idea without merit; it was a dangerous one.

Whereas some complex diseases may involve several changes in the DNA of a cell, like some cancers, and are believed to involve different steps and sometimes different factors, most human disease (even some cancers) can be thought of as involving a primary causal factor. Certainly this has been the case for most past epidemic disease for which we in time did learn the cause. Lewis Thomas, administrator, medical scientist, and author, explained in a speech I heard many years ago an idea I will paraphrase here: multifactorial is multi-ignorance. Most of the factors go away when we learn the real cause of a disease.

In an echo of one of the last lines in the film *Casablanca,* some people seemed prepared to "round up the usual suspects," placing in the dock many microbes known as troublemakers. For example, because Epstein-Barr virus can be a factor in abnormal B-cell growth, and because some AIDS patients were showing high antibodies to EBV and large lymph glands with B-cells, EBV was suggested as a cause of AIDS. But EBV is an old and ubiquitous herpes virus that of itself does not produce immune deficiency or a specific decline in the number of T4-cells—as occurs in AIDS. It would have to be an EBV variant to be the primary cause. But such variants of EBV have never been found.

Similar suggestions were made for *cytomegalovirus,* another ubiquitous human herpes virus, as well as for one type of hepatitis virus. An interesting possibility emerged from studies in New York City, where a cluster involving infection of one strain of adenovirus was found

My father, Francis

My mother, Louise

My wife, Mary Jane

Dr. Ludwik Gross, Cancer Research Unit, Veterans Administration Medical Center, New York City

Dr. Howard Temin, Professor of Oncology, University of Wisconsin. Nobel Prize–winning co-discoverer of reverse transcriptase

Dr. David Baltimore, President, Rockefeller University, New York City. Nobel Prize–winning co-discoverer of reverse transcriptase

Dr. Robert Ting, former Director of, now consultant to, Biotech Research Laboratories, Rockville, Maryland

Dr. William Blattner, Head of Viral Epidemiology, National Cancer Institute

Dr. M. G. Sarngadharan ("Sarang"), Director, Department of Cell Biology, Advanced Bioscience Laboratories, Inc., Kensington, Maryland

Dr. Flossie Wong-Staal, formerly with the Laboratory of Tumor Cell Biology, now Professor of Medicine and Biology, School of Medicine, University of California, San Diego

Dr. Robin Weiss, Senior Virologist and former Director, Chester Beatty Institute of Cancer Research, London

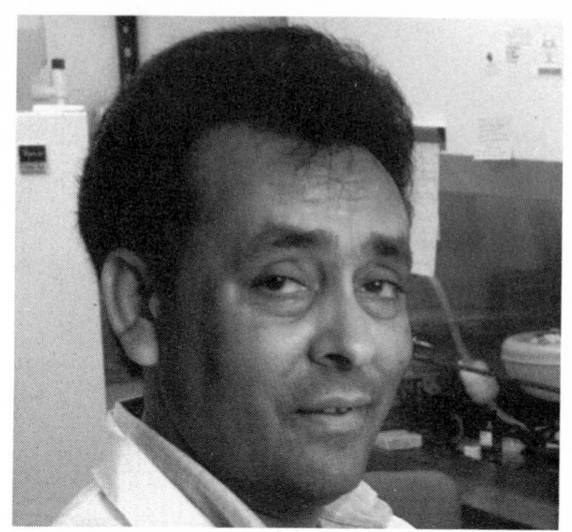

Dr. S. Zaki Salahuddin, formerly with the Laboratory of Tumor Cell Biology, now a professor in the Department of Medicine, University of Southern California

Dr. Marvin Reitz, Senior Investigator, Acting Head of Molecular Biology of Hematopoetic Cells, Laboratory of Tumor Cell Biology

Dr. Mitsuyaki Yoshida, formerly Senior Researcher at Cancer Institute of Tokyo, now in the Department of Cellular and Molecular Biology, Institute of Medical Science, the University of Tokyo

Dr. Marjorie Robert-Guroff, Senior Investigator, Laboratory of Tumor Cell Biology

Dr. Dani Bolognesi, Professor of Surgery, Duke University

Dr. Phillip D. Markham, Head of Cell Biology, Advanced BioScience Laboratories, Inc., Kensington, Maryland

Dr. William Haseltine, Chief of Division of Human Retrovirology, Dana-Farber Cancer Institute, Harvard University Medical School

Dr. Robert Huebner, one of the foremost U.S. virologists, formerly with NIH

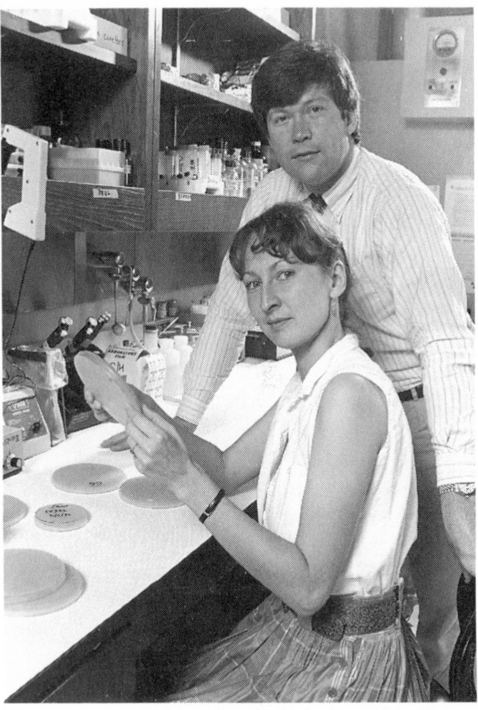

Drs. Beatrice H. Hahn and George M. Shaw, formerly with the Laboratory of Tumor Cell Biology, now in the Hematology/Oncology Department, School of Medicine, University of Alabama at Birmingham

Dr. V. S. Kalyanaraman (''Kaly''), Research Scientist, Advanced BioScience Laboratories, Inc., Kensington, Maryland

Dr. Françoise Barré-Sinoussi, Head of Retrovirus Biology Laboratory, Pasteur Institute, Paris

Dr. Luc Montagnier, Pasteur Institute, Paris; Dr. Myron ("Max") Essex, Chairman, Department of Cancer Biology, Harvard School of Public Health; and myself *(Courtesy of UPI/Bettmann)*

Dr. Jean-Claude Chermann, formerly with the Pasteur Institute, now Director of the Unit of Retrovirology Research, University of Luminy, Marseille

Dr. Jacques Leibowitch, Immunology Laboratory, Paris Public Hospitals

Visiting Senior Scientist Mikulas ("Mika") Popovic and Senior Technician Betsy Read-Connole, Laboratory of Tumor Cell Biology

Dr. William Jarrett, Professor of Veterinary Medicine, University of Glasgow, discoverer of the cat leukemia virus

Ersell Richardson, Senior Technician, Laboratory of Tumor Cell Biology

Dr. Daniel Zagury, Professor of Immunology, University of Paris

Dr. Howard Streicher, Clinical Scientist, Laboratory of Tumor Cell Biology

Dr. Isao Miyoshi, Dr. Kiyoshi Takatsuki, Mr. Armand Hammer, Yorio Hinuma and myself, at the awarding of the Armand Hammer Prize for Cancer Research, 1985

Dr. Genoveffa Franchini, Visiting Scientist, Laboratory of Tumor Cell Biology

among homosexuals. These DNA viruses exist in many strain forms. Some may infect T-cells. They usually cause relatively mild diseases. Adenoviruses are easily transmissible, however, a fact that could not be accommodated within our knowledge of the epidemiology of AIDS. Their easy transmissibility made possible a local epidemic among groups at high risk for AIDS, but an epidemic too different in too many ways from the AIDS we were looking at. Epidemiological follow-up studies showed that this was the case.

Several other reasonable ideas were suggested. Meanwhile some of the lay press and/or special interest groups had their own ideas, which varied depending on the readership. Among the surprising sources they blamed for the AIDS epidemic were the chemical amyl nitrate, used by some as a sexual stimulant; Agent Orange; the wrath of God and/or the work of the devil; African swine fever virus perpetuated by a CIA plot to kill pigs in Cuba, depriving the Cubans of food and thereby fomenting a revolution; syphilis; the deliberate laboratory creation of a virus by the U.S. government for germ warfare (for a while, the KGB tried to sell this one); the accidental laboratory creation of a virus by the Russians (this may have been the CIA's response to the KGB story); and a new agent created by the mixing of animal viruses by incompetent scientists.

From this baroque collection of ideas, we progressed to a state of impressionism, where almost any quick thought about AIDS by any proclaimed expert became newsworthy. Finally, we arrived at the apotheosis of the modern abstract—the view suggested and continually promoted by Peter Duesberg, a Berkeley scientist, that essentially nothing causes AIDS, certainly nothing *specific;* it's all just a matter of lifestyle. The obverse of this view, promoted by the same person, seems to be that almost everything can cause AIDS. Heroin, cocaine, antibiotics, various microbes—you name it. They can all do it.

Shortly after we first spoke, Max Essex began his own serological study of the AIDS virus, using HTLV-infected cells as a target for antibody reactivity. If people with AIDS are infected with a variant of an HTLV, they should have antibodies in their sera against the new virus, which will cross-react with the HTLVs. Essex would not attempt to isolate the virus itself.

Our earliest detections of reverse transcriptase activity in an AIDS patient dated to May 1982, with the cells of "EP." These cells were clearly positive for HTLV-1 proteins (or very related proteins) and we worked on this isolate for a number of months. Our next set of detections occurred between November 1982 and February 1983, with at least five positives, but none of these samples were as vigorous as EP and most showed at best low-level RT activity. Nonetheless, they appeared significant to us and they were negative for HTLV proteins; therefore they were suggestive of a new retrovirus. In February 1983, a number of new samples arrived, including a batch brought from Paris by Jacques Leibowitch. Several of these proved to be positive for RT activity. One of these RT-positive samples would be particularly important to our work. These were the cells from "CC," a Frenchman who had had an accident in Haiti requiring a blood transfusion. Now in Paris, he had been diagnosed as having AIDS.

Around the middle of February, shortly after the arrival of these new French specimens, Ersell Richardson began a primary culture of each of the samples sent by Leibowitch, including those of CC, using Interleukin-2 to stimulate their growth. She also established a co-culture with CC's cells, mixing his cells with human umbilical-cord blood cells.

On Day 15, Ersell found that the primary and the co-culture cells of CC were both positive for reverse transcriptase. On Day 42, the primary cells and the co-culture cells were tested for their reactivity to a protein associated with HTLV: the primary cells showed a 12 percent positive reactivity and the co-cultured cells a 48 percent reactivity—in both cases a significant amount. A later electron micrograph of the co-culture would come back showing HTLV-type virus particles but also aberrant viral particles with nuclei more condensed than those of an HTLV. All of us—Ersell, Popovic, and I—noted this peculiar result.

On Day 64, around mid-April 1983, Ersell noted in her log that the CC co-culture was still RT-positive and, more important, that both cultures were still going strong. Now we were quite excited to have cells from an AIDS patient in culture for several months, and we noted the important fact that the co-culture with umbilical-cord blood showed evidence of a strong cytopathic effect, the cell fusion phenomenon I mentioned earlier. Generally, once this phenomenon starts, the newly infected cells are not easy to keep growing. Yet CC's co-culture continued to grow. Looking at the physical appearance of the cells in culture, Mika pointed out that they did not resemble an HTLV cell culture; neither did their behavior correspond with that of the HTLVs in culture. There was clearly something different about CC's cells. But he did not yet know what it was.

Thus we had blood samples from people with AIDS in which we found retroviruses that were clearly related to the HTLVs, one of which, CC, grew in culture something like the HTLVs, although with more difficulty. Yet, totally unlike the HTLVs, this virus had the capacity to induce a very strong cytopathic effect when it was added to new target T4-cells. In addition, we had some samples from AIDS patients in which we found retroviruses not all detectably related to the HTLVs, and at that time many samples in which we found no evidence of a retrovirus at all.

By mid-April 1983, Max Essex informed me that he was ready to report his work in the scientific journals. So were we, and I informed Montagnier. We communicated frequently by telephone. I asked Essex to hold off sending his paper until Montagnier's paper could be submitted, so that all three might be published simultaneously. When *Science* asked for my comments on Montagnier's work, I strongly endorsed its publication. Montagnier had, however, omitted to write one important part of the paper, the abstract—the summary of one's findings that is included at the beginning of most scientific articles—and, given the pressures of time, I offered to write it for him. He agreed. Over the telephone, I read to him and to Chermann what I had written.

The results of the three groups' initial findings were published in the May 1983 issue of *Science.* My colleagues and I reported the isolation of an HTLV-related (or identical—we did not yet know which) retrovirus in one patient with AIDS (this was EP). We did not include the preliminary results on a new retrovirus (not cross-reactive with HTLVs) from a few samples obtained from AIDS patients because they were cultured for a very short time; in our view, the information was not sufficient for describing a new virus. We also published a survey of DNA from tissue samples from thirty-three patients with AIDS, in which we could find DNA sequences similar to HTLV-1 in only two. These two patients and EP as well would be the first of a fascinating phenomenon, but one that would very much complicate our findings. We also reported finding HTLV antigens in T-cells cultured from two other samples (these were the ones from Leibowitch, which included CC). We emphasized in our covering letter that naturally no etiological conclusions could be drawn from the results we had obtained, but "we wanted to bring attention to a new class of viruses (human retroviruses) as a possible cause of AIDS." If the virus was involved as a cause of AIDS, perhaps the low rate of detection of RT we had found could have been due to cell degradation from samples being killed by a cytopathic virus. It also might have been from too few cells being infected and very little virus being present. Our techniques might have frequently missed this virus, unlike in the HTLV-positive leukemias. After all, in AIDS patients the target cells are declining; in leukemia patients infected by the HTLVs, the target cells are expanding and thereby would be expected to yield more easily detectable virus.

Another important question, of course, was whether we were really detecting a minor variant, a mutant, of an HTLV. As I have discussed, a mutant would be interesting as a candidate cause, just like the minor variants of the cat leukemia virus, which can cause an AIDS-like disease. The alternative was that we had found immortalizing HTLV-1 or HTLV-2 themselves. If so, these could only be co-factors or irrelevant co-infections since they are older infections of humans and AIDS is a new epidemic. They would thus be of little or no relevance in finding the primary cause of AIDS, much like many other infections in AIDS patients, but still of some interest and usefulness.

The paper by Max Essex showed that with coded sera (a "blind" study), about 35 percent of AIDS sera tested positive for some kind of reactivity with HTLV-1–infected cells. If anyone seemed to be coming close to proving anything at that time, it was Max, but what he seemed to have demonstrated—that a virus closely related to HTLV-1 was involved with AIDS—turned out not to be true.

The paper by Montagnier and co-workers described their findings from a patient with lymph gland enlargement (lymphadenopathy) but without immune deficiency. This was the patient BRU whose cells kept dying, but now they reported the cells as being cultured in human umbilical-cord blood T4-cells. When I first read their paper, I was intrigued by its evidence that this new virus was likely to be an HTLV-1 variant, perhaps a substantial variant but still a virus of this family.

What made me think this new retrovirus would be a member of the HTLV family? The conditions they'd used for their RT assays—the buffer, the pH, and so on—were ideal for HTLV-1. They were exactly those we had used and described for HTLV-1 in 1980. And the major structural protein of their virus was smaller than those found in almost all animal retroviruses but identical in size to the comparable structural protein of HTLV-1 and HTLV-2. Also, they noted that their virus infected T-cells (as did HTLVs), and they found that the serum from their patient reacted with HTLV-1–infected cells. They also stated that electron microscopy showed "characteristic . . . C-type budding at the plasma membrane," similar to HTLV-1 and HTLV-2. Some of their findings, such as the description of their virus as C type and the cross-reactivity with HTLV-1, proved to be wrong.

Nonetheless, what they had clearly differed in important ways from what we knew about the HTLV family, at least in regard to HTLV-1 and HTLV-2. For example, to keep the virus going in culture, they had had to regularly infect T-cells from a new source of baby blood, as we now understand, because the original source and its target cells would quickly be destroyed. The electron micrographic pictures they displayed also looked different from those of HTLV-1 or HTLV-2. Further, reagents I sent them that were specific to HTLVs could distinguish their virus from the HTLVs. The overall evidence suggested to me that this was a new human retrovirus, one likely related to the HTLVs. Presumably, this might be what Essex was finding in his tests of sera for antibodies. Those positives might indeed be detections of this new retrovirus. Could this also be the same as CC? Or could it be those retroviruses we detected that did not cross-react with the HTLVs? Such were the questions on my mind.

Shortly after Montagnier's article appeared, however, four very experienced electron microscopists began to question their published report: Jack Dalton, for decades the leading electron microscopist for retroviruses in the United States; John Moloney (discoverer of the mouse Moloney leukemia retrovirus and the mouse Moloney sarcoma retrovirus and former head of the NCI Virus Cancer Program); Robert F. Zeigel, an electron microscopist at Roswell Park Institute in Buffalo; and Françoise Hageneau, the leading electron microscopist for retroviruses in France. Uniformly, they believed the report failed to show a retrovirus; instead most of them thought it was an arenavirus.* And they communicated their opinions strongly.

The Pasteur group was, in fact, right. They had found a retrovirus—

*Like retroviruses, *arenaviruses* are enveloped, budding RNA viruses, but there the similarity ends. Many other properties distinguish them from retroviruses.

indeed, a new one—and the French group, to their credit, would stay with what they had through the criticism that followed. Even I was less certain about their findings than I had been initially, but many scientists I spoke with at the time simply did not believe they had found a retrovirus at all, let alone a novel one. The problem, of course, was that their data were incomplete. The electron micrographs of the virus structure were not very clear, and there was no evidence yet linking the virus to AIDS. Having had some experience with human retroviruses, both successful and failed attempts, we knew what the French group had to do next. They had to come back not simply with better micrographs, but with stronger data, with every counterpossibility fully explored, with enough virus to work with and to give out to others, and with convincing evidence demonstrating whether or not it was linked to AIDS. These were the same tasks that we had set for our earlier work with the HTLVs and leukemias caused by them, and that we would set for ourselves for AIDS. Before we would publish another paper, we would have many, if not all, of these biologic problems solved. Still, the Pasteur's reverse transcriptase data did seem clear-cut, at least with regard to the fact that their isolate did contain evidence of a retrovirus, and the virus surely was novel.

CC's cells continued to give interesting but complex results. Without any fanfare, Ersell Richardson marks in her log, "Cells poor," on May 16, 1983, of the co-culture of CC and, "Cells are dying," on May 31, 1983. This had never before been our experience with a pure HTLV. Something inexplicable was going on. Though they reacted positively to HTLV reagents and, like HTLV particles, easily infected umbilical-cord blood cells, unlike HTLV-infected cells, they were experiencing growth crises. Normally, once HTLV-1 and HTLV-2 immortalize their target T-cells, they have no growth crises.

Was CC the sought-for HTLV variant, the cause of AIDS? Or were viruses from CC's cells of two types—one an HTLV itself, the other a novel retrovirus? A picture in a jigsaw puzzle was taking shape as more pieces began to fit, but the whole was not yet visible. We decided that George Shaw in Flossie's group should try to pull out the viruses from CC's cultures and molecularly characterize them. We awaited the results.

From May to June 1983, I decided to expand our AIDS research efforts. I asked Phil Markham and Zaki Salahuddin to join in the cell culturing and virus isolation parallel to, but independent of, the work of Popovic (and of the technicians Betsy Read-Connole and Ersell Richardson). I had already asked Marjorie Robert-Guroff to check out the AIDS patients' sera carefully for antibodies to HTLVs, even weakly reactive ones.* Would we get results comparable to those found by Essex using our methods, which employed not virus-infected cells as the test for antibody reactivity but purified virus itself?

Her findings would soon surprise us. In contrast to Essex's results (30 to 40 percent seropositive), only about 10 percent of people with AIDS tested positive for antibodies reactive with HTLV-1 or HTLV-2 (the exact percentage depends on the region, the type of risk group the sample comes from, and the time of sampling).

As interesting, Marjorie's score did not increase if she "relaxed" the conditions of her tests, which surely should have picked up any slight variants of an HTLV. At just about this time, we obtained even more disturbing news from Ed Gelman's work. The few cases my colleagues and I had previously found positive for HTLV-1 nucleotide sequences, as well as our first isolate, both published in the May 1983 *Science* article, were really HTLV-1 itself and *not variants.*

Veffa Franchini, who had joined our lab from Italy just a few years earlier, remembers this period of late summer to early fall 1983: "We were still thinking that an HTLV-1 variant—but now, to a much greater extent, not a minor variant—could be the cause of AIDS and there were many reasons to believe it was, but there were also other reasons not to believe that. It was the time in which Marjorie did have 10 percent positivity for HTLV-1 antibodies in patients with AIDS. And I remember that we had the lab meeting, staff meeting, and we had a round table at which everybody was asking at the end, 'Do you believe or don't you believe that an HTLV-1 variant is causing AIDS?' " And she notes that I went around the table just asking everybody, "What do you think? What do you think?" She remembers "that the majority of people really believed that an HTLV-1 slight variant was not the cause of AIDS. The majority thought that it was another virus. Most of it, the negative votes, was because of the

*Though, as I mentioned, we also had by then several detections of a new retrovirus, we could not really test sera for antibodies reactive to it very well at this period because we had not yet been able to grow the virus properly. Any results would likely be misleading. As we will soon see, CC had not only HTLV-1 but also the new (AIDS) retrovirus. We did not pursue use of CC for antibody testing of sera from AIDS patients because of the confusion we had about what kinds of virus or viruses were present in CC.

Marjorie paper." Robert-Guroff would present this data at the upcoming Cold Spring Harbor meeting, but no publication in a peer-reviewed journal would be made on retroviruses or AIDS by us or by the Pasteur group for the remainder of 1983.

By the summer of 1983, we all sensed that if we were to find the cause of AIDS, we had to move in on it fast. Not yet knowing what caused the disease, we could not offer any definitive answers about how quickly it would spread. Though epidemiological reports were starting to suggest that major symptoms might not appear for many years after exposure, we did not yet have a blood test and could not know how many people had actually been exposed and were now walking around with a time bomb in their blood. The sinister specter of not being able to develop a blood test until virtually everyone had been infected loomed before us.

But our cell culturing was now speeding up. Phil Markham and Zaki Salahuddin proceeded to remove as a factor the time blood spent in transit. Both became as often as possible their own mini-courier service, "beating the bushes," as Phil remembers it, "for new sources of materials," especially when an interesting situation came up. For instance, Phil recalls getting into one of "those four-seater aircraft to go up to Yale, taking with me a bag containing media and separation tubes. I had been told there was sera available from a symptomatic baby and an asymptomatic mother. I carried the separated cells back with me the same evening."

When he later examined them, the cells from both the symptomatic baby and the asymptomatic mother were positive for reverse transcriptase. That was one of the earliest detections Phil and Zaki had and, as Phil states, it "provided probably the first indication that the AIDS virus could, in fact, be transmitted either by IV drug abusers [the mother was a drug abuser] and/or by heterosexual routes, and also to babies of infected mothers."

Salahuddin began to travel to New York every Friday to Dr. Bijan Safai's clinic at Sloan-Kettering Memorial Hospital to collect blood from his AIDS patients. Zaki reminded me how he used "to set up experiments right there in Bijan's lab and bring them over to our lab for further analyses. But invariably," he noted, "we had the same problem the Paris group was having. These cells would transiently express reverse transcriptase and shortly thereafter die. We were waiting for a stable virus producer."

At our July 18, 1983, AIDS Task Force Meeting, an ad hoc group of scientists that included myself, Bill Blattner, Dani Bolognesi, Wade Parks, and Max Essex listened to Luc Montagnier, whom I had invited, present some material on his AIDS work. One slide, used in his presentation on virus isolation attempts from AIDS and ARC* patients, contained virus-like particles. Matthew Gonda, an NIH electron microscopist, joined us. He recalls:

"One of the images in the slide caught my attention. It was of an extracellular viruslike particle that had a bar- or cone-shaped core. . . . It was mentioned that the particles might be an arenavirus or possibly a type-C virus. Because of my background in electron microscopy, I was asked to review several additional micrographs that Dr. Montagnier had brought with him. It was clear to me from these photographs that the viruslike particles were not arenaviruses or type-C oncoviruses, but were very similar to images of equine infectious anemia virus (EIAV), a virus whose structure I had studied and written about several years earlier. I told the audience that EIAV was *a member of the lentivirus subfamily of retroviruses.* Dr. Montagnier said that he had never heard of the lentiviruses. I explained that most of the members of this group infect cells of, and cause perturbances to, the immune system. The diseases are slow in onset but chronic and progressive, often leading to death. Some of the lentiviruses, visna and caprine encephalitis viruses, also caused neurologic disorders in their hosts."

Gonda was right. He was the first person to recognize that the new retrovirus—what we would soon show was the AIDS virus—is in the lentivirus subfamily of retroviruses.

Later at this same meeting, Montagnier and I talked. In all our previous discussions, we had agreed that the molecular analyses of his isolate (later called LAV) would be done by us in a collaborative effort with their lab. Both of us understood what a collaboration meant. Any of their isolates sent to us would be given the same treatment as our own isolates: whatever we learned to do with one we would do with another, and the results would be published as a collaborative effort under the names of members of both labs. Now Montagnier pulled me aside to tell me that the collaboration was off. Under pressure that

*ARC = AIDS-Related Complex, a term coined, I believe, by the CDC to describe pre-AIDS signs.

"it all must be done in the Hexagon," he said he had to change the plans. I had a good idea what he meant by referring to the geometrical shape of France.*

In August 1983, I wrote the director of NCI, Vince DeVita, that my notion of the cause of AIDS was still that it was a human T lymphotropic retrovirus, still related to HTLV-1 and HTLV-2, but clearly differing from them in major ways. Examining the cells of CC, Ersell made two further notations of cells in crisis, but the cells remained in culture throughout August. One month earlier, in an article in the *Journal of the American Medical Association,* based on an interview I had given in June 1983, I wrote that there was now enough reason to believe the AIDS virus was cytopathic because there was clear evidence that "the target cells . . . disappear once they are infected."† We were surely aware that any retrovirus involved in the cause of AIDS would be substantially different from the known HTLVs. Yet the bulk of evidence, though conflicting and confusing, suggested to me that they would be related.

Back in April 1983, as part of our collaborative effort with the French, we had received a sample of DNA from LAV-infected cells. We had not been able to detect any DNA sequences related to HTLV-1 in this sample; nor did we hear back from the French, to whom we had given an HTLV-1 probe, that they had either. This was a bit surprising because in their first published article they had reported an immune cross-reaction with HTLV-1. Of course, without a probe specific to viruses other than the HTLVs, we could not tell whether DNA from another virus or viruses was present.

Most of our own tissue samples, even those that were reverse transcriptase–positive, did not show an HTLV-1 sequence in their DNA, either. But, as important, some did. Not many, but enough to be of interest, and we wondered what this might mean. Clearly the patients from whom these cells had been taken had AIDS. Was HTLV-1 itself a passenger, an opportunistic infection, or was it part of the disease?

More so than with any other isolate, the cells of patient CC would help us answer this question. As we became more convinced that a close variant of HTLV-1 was not the cause of the disease and as the

*In short, from July 1983 on, we had approval to experiment with LAV particles but not clear permission to publish any results with LAV.

†To my knowledge, this was the first time the virus was said to kill its target T-cells.

molecular analysis of the HTLV-1 from CC and EP became available, we also knew that the cytopathic effect observed in the CC cultures was not due to a variant of HTLV-1, because the virus was indeed HTLV-1 itself, which undoubtedly was responsible for inducing the permanent growth of CC's cells, albeit with difficulty. By the summer of 1983, we were driven to the conclusion that the virus we were looking for would not only be a third human retrovirus but that it would differ substantively from the HTLVs.

In July, Montagnier, at our request, had sent us supernatant fluid* containing virus particles from their LAV isolate. Mika did some work with it but there was not enough virus to carry out meaningful experiments. He requested a second sample. But by this time, the work of Markham and Salahuddin was already moving rapidly and producing a variety of extremely interesting findings. In June 1983, for example, they had separated plasma, that part of the blood that is cell-free, from blood taken from an AIDS patient. If virus were reproducing in body cells, the new virus particles would be released into the plasma. They showed that the plasma was indeed positive for virus production, and the virus particles were HTLV-negative. In another experiment they took cells from the bone marrow of an AIDS patient to look for evidence of virus production there. Again, they found what they were looking for.

We were now all preparing for the Cold Spring Harbor Laboratory meeting in honor of Mary Lasker, which would take place in September, in the town of Cold Spring Harbor on the North Shore of Long Island. This was a wonderful place for a meeting—a lab with an excellent reputation for top-notch work in one of the most beautiful physical settings in the East. I had been there on about a dozen previous occasions. But this was the first meeting to be held there on human retroviruses; in that respect it was almost historic. I was one of the few meeting organizers. Although leukemia and the HTLVs were the major focus, there would also be much talk about AIDS.

By the time of the meeting, Markham and Salahuddin had more than twenty detections and/or short-term isolates of retroviruses clearly distinct from HTLV-1 and HTLV-2.† On co-culturing, eight

*When virus is released from cells into the media cells are grown in, the fluid containing virus but without the cells can loosely be called *supernatant.*

†I call here an *isolate* any virus particle that is transmitted to a target cell, even if infection is evidenced for only a short time.

were also shown to be cytopathic. Were they all of one virus type, or several? Virtually none of these isolates was sent out for electron microscopy because Markham and Salahuddin were still waiting for a high-level burst of virus release. Meanwhile, Ersell Richardson had made another eight detections of this new retrovirus.

To me, the questions were obvious. Were Phil, Zaki, and Ersell detecting the same virus as LAV, or a different one? Were their cultures somehow related to the CC culture of Popovic and Richardson, even if CC was clearly closely related to the HTLV-1? Or were their cultures more like Popovic's other early detections which, unlike CC, were not detectably cross-reactive with the HTLVs? Or could the T4 cultured cells infected with one retrovirus contain more than an HTLV and the new (AIDS?) retrovirus?

More important, were they the key to AIDS?

Though neither our group nor the Pasteur group would publish any more findings in 1983, much would happen in the final months of that year. By the following spring we would be publishing the answers to most of these questions.

It may be useful to distill the story told in this chapter. Throughout 1982 and into mid-1983, some of our findings suggested that the causative agent of AIDS was a retrovirus closely related to HTLV-1. But other findings suggested it was not.

Supporting its being an HTLV variant were the following findings: (1) We were able to make several isolates of HTLV-1–like viruses from AIDS patients. (2) Tests by Essex for antibodies in the sera of patients with AIDS yielded better results than did similar tests by Montagnier, even though Essex tested for antibodies against HTLV-1 and Montagnier used the new retrovirus, LAV. (3) In his 1983 paper, Montagnier reported that sera from his first patient reacted strongly with HTLV-1–infected cells, and he identified LAV as a type-C virus, putting it in the family of the HTLVs. (4) Our most promising culture, "CC," also showed signs of HTLV-1 proteins. This isolate seemed to be the best candidate for an AIDS retrovirus because it killed some target T-cells in laboratory tests, and because electron micrographs of CC showed the presence of strange-looking virus particles that fit, given our ideas at the time, with a finding of a variant of the HTLVs.

The results favoring a retrovirus not clearly related to the HTLVs were as follows: (1) Beginning in November 1982, we began finding detections of retroviruses that did not cross-react with the HTLVs. (2) By the summer of 1983, several of these had been transmitted to new target T-cells and were grown in culture at least temporarily. They showed cytopathic effects on these target T-cells, and these isolates, when tested by one or two specific antibody tests, were not detectably related to the HTLVs. (By the end of the summer we had about twenty of these detections or isolates.) (3) By late summer/early fall 1983, the results with LAV were clarified by Chermann and Montagnier. They no longer described LAV as a type-C virus, and they recognized that their earlier showing of a cross-reaction of their patients' sera with HTLV-1 was erroneous. (4) By late summer 1983, our group knew that the HTLV-1–like viruses we had isolated earlier were not the sought-for variant but immortalizing HTLV-1 itself. As such, they could accompany, but not be, the real causative agent. (5) Our own serological results (produced by Marjorie Robert-Guroff) showed that the sera of 10 percent or fewer of AIDS patients contained antibodies reactive to the HTLVs. This also correlated with what others were seeing in our laboratory: 10 percent or fewer of our isolates reacted with HTLV reagents and/or yielded HTLVs by molecular probing. Whereas earlier we had wondered whether this was because our test was simply not sensitive enough (given the much higher correlation Max Essex had obtained), we now believed that 10 percent was the upper limit—that is, only 10 percent or fewer of AIDS patients were also infected with an immortalizing HTLV.

These last two findings (4 and 5) led us to realize that Essex's interpretation of his serologic findings had to be wrong. (He was picking up a cell protein that somehow reacted with some antibodies in the sera of 35 percent of AIDS patients.) We now also considered the possibility that Montagnier might have gotten a low response using LAV and AIDS patients' sera—he got only 20 percent reactivity—because of technical problems rather than because he had the wrong agent. (6) Last, the notion that some AIDS patients bore a double infection—HTLV-1 or HTLV-2 plus the new retrovirus that was likely the cause of AIDS—began to gather adherents. The only explanation for CC was that it contained two viruses: an immortalizing HTLV and the AIDS retrovirus. This, in turn, led us, especially Mika Popovic, to conclude that Montagnier had been wrong in his statements, made repeatedly, both in public and in private, that the new kind of retrovirus could not be grown in a cell line. CC was it.

In this new frame of mind, our group headed for the Cold Spring Harbor meeting in September 1983. But if our ideas were now firming up, we still did not have laboratory results that clearly linked this new cytopathic, probably LAV-like, retrovirus to the cause of AIDS.

9

Breaking Through: "We Know How to Work with This Kind of Virus"

In some ways all scientific gatherings are alike, and in some ways very different. None are laid back; almost all have their special and occasionally dramatic moments; most are basically collegial; some are strongly competitive. I remember a scientific meeting early in my career where I was one of those presenting a paper. What I spoke about is of little importance now. After I had finished, Sol Spiegelman got up to question me. Had I considered this? Had I ruled out that? How did I know this? Why did I not think it was that? On and on and on and on. I didn't have good answers to his questions. I should have, but I didn't.

Afterward, Bob Ting tried to console me, telling me that Spiegelman, known as a very tough guy, wouldn't have taken the trouble to point out what I had left uncovered unless he believed in my work. Take it in a positive way, Ting told me.

Back then, I was still young and it was not easy to shrug off the sting of criticism. Over time, however, I came to know Spiegelman as a friend and to value that first experience with him. It had taught me an important lesson. If science is one part inspiration (the ability to think originally, to come up with creative hypotheses), and a second part tenaciousness (the willingness to stick to your ideas while others reject and denounce them), the third part is the proof, the long, hard benchwork Thomas Edison must have been thinking of when he described genius as 98 percent perspiration.

Story after story of scientific discovery has followed this pattern. Howard Temin, for instance, believed in the RNA → DNA path for a retrovirus for years before the rest of the scientific community accepted it; to his credit, he stayed with it and he and David Baltimore, who worked independently of Temin, finally forced the scientific community to accept one of the truths it was lucky to have learned when it faced the AIDS crisis years later.

I hope it is not presumptuous of me to say that I think we went through much the same process in our own lab. We believed in the human retrovirus and stuck with our beliefs throughout our own remarkable journey of discovery, apparent confirmation, bitter disappointment, and then conclusive evidence. But until we were able to lay out our proof before the scientific community and invite our research colleagues to show, if any could, why we had not done what we claimed to have done, no credit came our way. This is the way of science. Most of us would not want it any other way.

Knowing what lies ahead when we present findings to our peers shapes the thinking of all of us in the lab, especially me. My staff meetings are held almost weekly. Although informal and with no set agenda, they can be grueling sessions for anyone making a claim for his or her work. When asked recently what makes our lab distinctive, Prem Sarin began by noting the fact that we have "a superaggressive boss." Yes, I have withered young postdocs who told of their findings without having done the work necessary to defend their results. I admit that with tough questioning I have at times, unfortunately, hurt the feelings of some of my staff, especially those not used to my ways. I do it not because I am on an ego trip but because I believe in the value of vigorous debate. It is almost instinctive.

I see it as part of the scientific process, part of the hard-knocks survival selection process that evolves a hardier truth. In the early years, before AIDS changed our perspectives about time, we would hold an annual lab meeting in a park. For a couple of days we would review what we had done over the past year and map out where we should be going in the next year. Patting each other on the back for our past achievements was great fun. But it was not what the meetings were really about. They were about charging each other up by challenging each other's thinking.

Prem Sarin remembers those meetings as a "a free-for-all. Everybody was young and eager to ask questions. And you felt free to ask whatever you wanted to. Every time you got up, you would expect to be questioned. Ninety percent of the time it would be against you. You didn't do this or you didn't do that. Have you looked at it this way? Have you looked at all? The questioning was vigorous, very vigorous, but that was good in a way because this is what opens an area for you to go, this is what forces you to move your energy in that direction."

This kind of questioning, which by now is part of the style of our lab and probably of many productive labs, would become the second source of conflict between Montagnier and me. The first he had initiated with his decision not to collaborate, after I thought we had months earlier agreed to do so and had provided him with some technical help.

The Cold Spring Harbor Conference is an intense meeting of senior scientists and very dedicated young people. Years ago, senior people regularly spoke, often to present speculative findings. The meetings had a different tone then, more reflective, a bit less businesslike. At more recent meetings, it is generally the young postdocs who do most of the presentations. Although senior people sometimes speak, more often they do the questioning—which is always vigorous and sometimes merciless. Stories about these meetings are legendary.

"If you talk with any young postdoc about the Cold Spring Harbor meeting," recalls Veffa Franchini, "they are terrified. In fact, Antoine Gessain, a postdoctoral Fellow from Lyon, France, told me a story about waiting for his turn to speak when this guy, another speaker, fainted. It is really very high pressure for young people. You're expected to go there, but not unprepared. The interrogator can ask you whatever he wants. And that's always done. The questioning can get very nasty. But it's the best meeting for young people in biology, especially in human virology."

Flossie Wong-Staal, who came to the lab as a postdoctoral Fellow and worked in the molecular genetics division of the lab for close to sixteen years (and as its head for more than half those years), tells a similar story: "It's a wonderful meeting for young postdocs because you talk almost nothing but science for the whole meeting. It's collegial but it's also a very competitive atmosphere."

Flossie also recalls that "the people from our lab were often doubly nervous about the meetings. We always felt at a bit of a disadvantage because we'd been working on human retroviruses for so many years, and this was not very popular in that period. In fact, in the early days, they always put the human retrovirus session last—on Sunday morning, the last event of the last day of a long and grueling meeting, so usually by the time our session came along, you'd have maybe thirty or fifty people, stragglers, that's all that was left. So we were very sensitive in terms of our data being impeccable. We did not want to be knocked down in public, we wanted to be sure everything was right, that we had thought about where we might be vulnerable before someone else pointed this out to us. I don't think any one of those who questioned us was mean, but the overall atmosphere, well. . . ."

By 1983, of course, we were no longer trying to prove ourselves in the same way. As an indication of just how much things had changed, the upcoming Cold Spring Harbor meeting was on human retroviruses, and I was one of the meeting's organizers. Most of the meeting was to be devoted to the viruses we had discovered and helped link to the cause of some human leukemias—the HTLVs. But AIDS was on my mind, and my own thoughts were frankly less on how other scientists attending the meeting would view our findings than on the state of flux of our own ideas about the cause of AIDS. The lab was now convinced that the etiological agent of AIDS could not be just a minor variant of either HTLV-1 or -2. Marjorie Robert-Guroff would report on her study of sera in AIDS patients for antibodies to HTLV, showing only a 10 percent seroreactivity.

Her findings were not far off what we had reported in our first *Science* articles—that two out of thirty-three DNA samples from AIDS patients' cells were positive by molecular probing with an HTLV probe.* Also, Ed Gelman and Flossie Wong-Staal had given us the results of their molecular analyses of these two HTLV-positive samples. They reported that the two cases were not HTLV variants but HTLV-1 itself.†

*Although the two positives out of thirty-three specimens examined in 1982 and early 1983 were already clear negative results for HTLV in AIDS, it was conceivable that too few cells were infected to be detected by the molecular probing techniques then available. Thus, the question remained open.

†As I have indicated elsewhere, later we would learn that none of those HTLV-positive AIDS cases were ever minor variants of HTLV-1 or HTLV-2. Instead, these people were doubly infected with either HTLV-1 or HTLV-2 *and* the AIDS retrovirus.

These new findings undermined our first thesis—that the AIDS virus was a minor variant of HTLV-1 or HTLV-2—but they also gave us our first real clue to our future direction. We had now months of experience with the cells of the AIDS patient CC, the young Frenchman who had been infected by a blood transfusion in Haiti. Remember, not only did his cultured T4-cells continue to grow and contain retroviruses highly related to the HTLVs, but there were two peculiarities about the viruses produced by CC's cells: they also contained some viruses different in structure from the HTLVs (discovered by electron microscopy); and when we transmitted these viruses to target cells, they usually induced cell damage—the cytopathic effect. Moreover, CC's cells themselves grew with difficulty—none of this was consistent with the HTLVs.

These observations now suggested that CC contained an HTLV itself, *but* that the "aberrant" particles were not altered forms of the same virus but a second retrovirus in the same cells: that is, CC was doubly infected by his Haitian blood transfusion. (In Haiti HTLV-1 was, and is still, endemic and AIDS was then becoming epidemic.) Could CC's growth crises, multinucleated cells, and aberrant virus particles spotted by electron microscopy be the work of one virus (perhaps the virus that is the cause of AIDS), and the long-term survivability the work of another, a true HTLV? In short, was CC infected with two very different kinds of retrovirus? If so, was CC trying to give us a clue about how to solve the long-term growth problem of AIDS-infected cells?

At Cold Spring Harbor, we presented the HTLV-1 serologic data but would say little of our new thinking or our new findings. Though Robert-Guroff presented the serum antibody HTLV-negative data in detail, and we did briefly discuss and indicate its implications (against a closely related HTLV retrovirus), it was at this point only speculation. I had decided after the publication of our first paper that we would make no further publications or announcements until we had everything, or at least most of it, figured out. So we also said nothing yet about the twenty or so new detections of the novel retrovirus.

Montagnier's group took a different approach. They did report their latest findings at Cold Spring Harbor, contradictions and all. Montagnier did not conclude that their virus was the cause of AIDS. His style of presentation was matter of fact. But there was a contrast between his steadfast commitment to the major significance of their isolate and the casual way he seemed to be responding to the responsibility to lock up this finding with hard proof.

Francis Crick, in his recent book *What Mad Pursuit,* describes how J. D. Bernal, the founder of X-ray diffraction studies of protein crystals, would politely critique a colleague's talk, prefacing his points with compliments. Crick explains how this habit surprised him at first, but he eventually learned that "if you have something [critical] to say . . . it is better to say it firmly but nicely and preface it with praise of any good aspects of it." Crick saw himself as too impatient and impulsive and inclined to express himself "too briskly and in too devastating a manner." This was a warning I should have heeded.

Montagnier's presentation had new information of interest. He began by noting that in five patients with swollen lymph nodes and in three diagnosed as having AIDS, he now had evidence of a virus resembling LAV. He re-presented his earlier work on the structural differences between LAV and the HTLVs. He again pointed out that his group could not make antibodies to LAV in animals and, oddly, that infected people did not appear to make antibodies to the virus envelope.

He now had some curious serological findings to report: he had found antibodies to LAV in 60 percent of the people with lymphadenopathy but in *only 20 percent of those with AIDS,* a point that, I think, brought him to emphasize that LAV might cause the lymph node abnormalities, and that *many other retroviruses, perhaps also LAV and the HTLVs or variants of them, might be causing AIDS.* These are the discussions he made in his written version of the presentation. He also noted again that LAV could not be grown in a cell line.

Instead of first acknowledging, à la Bernal, the importance of his new information and commenting on the value of his contributions, I proceeded at once to ask for more details about some of Montagnier's assertions, challenging him with questions.

His electron microscopic results were not convincing, and the retroviral nature of his particles depended on the definitiveness of his assays for reverse transcriptase. What were the details of the assays? How did he propose to defend his assay findings if they were not as complete as perhaps they should have been? If no specific reagents were available, how could he tell whether the new detections were all of the same virus type? How could he be sure they were not contaminants with animal retroviruses if he had not done the kinds of experiments we had done with HTLV-1 before the scientific community accepted our discovery of it? Consistent with the fact that he found antibodies to LAV in 60 percent of people with lymphadenopathy and only 20 percent of those with AIDS (but not, in my mind, with the logic of probabilities), did he really believe LAV was involved early but not late in the disease?

Montagnier was clearly rattled by the questions and seemed particularly disturbed that they had come from me. We are not alike in our styles, as people or as scientists. He is quiet, almost formal, holding his own counsel when competing ideas are being presented. If he speaks at all, it is usually simply to ask a question. I love the rough-and-tumble of intellectual debate and usually welcome attacks on my own ideas (though I admit not always, and depending on the source), even though I know that at the moment I will be uncomfortable with them. If the attack has merit, I will join in myself, adding to the critic's points others that have bothered me about my thesis.

If Montagnier had been better prepared to deal with my questions, he might have backed me off. Instead, he began to theorize that AIDS was caused by many different kinds of retroviruses, including HTLV-1 and HTLV-2 variants, LAV, and possibly two new viruses called IDAV-1 and IDAV-2 (the *ID* standing for "immune deficiency"), as well as retroviruses yet to be discovered. He discussed this point more extensively in the paper he later published from this meeting. Here and elsewhere he also implied that LAV might be causing the lymphadenopathy and another retrovirus (or other nonspecific factors that can influence the immune system) might be causing AIDS.

This was not an idea I was very fond of. I believed it was obvious that one new agent would have to be the cause of the disease—allowing, of course, for opportunistic infections that will invade as immunity is destroyed and for the possibility that some factors *might* sometimes hasten the progression of the disease, which is caused by the one agent. It was Chermann of the Pasteur group who also always believed in one specific cause, and felt it was the LAV-type retrovirus from early on. I think he felt their inability to link it to AIDS was due only to their technical difficulties.*

*In fact, to the best of my knowledge, Chermann was the first to believe this, remaining consistent and always open in his view even though he did not have convincing data at the time. I did not have too many chances to talk with Barré-Sinoussi; my guess is that she was of the same view as Chermann.

This, too, may help explain my impatience with Montagnier's answers that day. I did not fully understand the lack of technology available to Montagnier, so I did not consider that he might have failed to do the many things he might have done before coming forward with new claims because he simply did not have access to the proper resources or experience. I assumed too much. Perhaps, had he not chosen to sever the collaboration between us, we could have been of more help to him. My reasonable assumption was that he would not have turned the collaboration down if he was incapable of replacing those resources.

But in the world outside the lab, perception is reality and I have often thought about how much simpler life might have been over the next few years had I not led the charge that day. I was, of course, in the mode characteristic of Cold Spring Harbor meetings, primarily posing aloud questions colleagues had raised in private throughout the conference. Further, though we were the hosts and he a guest, I knew that seminars at the Pasteur Institute, led by François Jacob, its president, were often filled with critical questioning, even of foreign guests, though these seminars were admittedly much smaller than the Cold Spring Harbor meeting.

Still, I have come increasingly to regret that the tone or spirit of my questioning that day was too aggressive and therefore misunderstood. It widened the growing chasm between the two labs that were making the greatest contributions toward eventually controlling AIDS.

Later the rift was seen in international terms. Much of the French lay press would support Montagnier. The American press would start off accepting the other side's view of things, and suspect me simply because I am government-based. American coverage was sometimes based on the assumption that the French had fairly and disinterestedly put forward their case, while the Americans had been driven by motives of greed, aggressiveness, and competition. The French press is known for its boosterism, the American press for its skepticism, which it sometimes carries to an extreme. Montagnier thus became the maxi-darling of the French press and, where and when he was known here, it would be as the mini-darling of the American press. I became the outsider in the French press and a villain in some quarters of the American press. Though it often frustrated me, this all had to do with national style and character, and I soon accepted that it was much larger than the dispute and clearly beyond my power to change.

Despite the friction between Montagnier and me, our two labs continued to work with each other, following the only responsible course of action available to us.

One last event of note took place at that Cold Spring Harbor meeting. European, Japanese, and American investigators signed an informal agreement stating that any new human retroviruses would be called by number in the order of their discovery if the target cells were chiefly T lymphocytes: HTLV-1, -2, -3, -4, and so on. Thus, if this was a new virus infecting T-cells, its name would be HTLV-3. Different strains of any one kind (if differences within a type of HTLV were found) were to be distinguished by the initials of patients or some other subscript code.

Although we did not know it as we left the conference and headed back to Bethesda, the proof I had pushed for from Montagnier would be shortly forthcoming—from our own lab. I had requested that Phil Markham and Zaki Salahuddin focus on getting as many reproducible detections as possible of the new type of retrovirus from AIDS patients and risk groups. If these were all one virus type, this would greatly help establish a causal relationship. They would also include many samples from healthy people not in risk groups as controls. In contrast, Mika would focus on trying to solve the problem of producing the new kind of retrovirus in pure form (CC would not do for this purpose because of the HTLV-1 presence) in a permanently growing cell line. Mika had formulated and given to our technician Betsy Read-Connole methodological plans for approaching the cell-line problem, and ironically the work Phil and Zaki had been doing would shortly (in October 1983) result in their first long-lived isolate.

In early September, before the meeting, in thinking about CC's cells, Mika had made a number of important decisions. The first was to try to get a strong isolate started by employing a novel technique. Instead of just culturing each patient's cells separately, as he had been doing for some time now, he decided that when he could not get a sufficient volume of culture fluids from the cells of one patient, he would pool together different samples, using both concentrated culture fluids and cells for infection. He suggested a number of different target cells, including permanent cell lines.

But the decision to try hard to transfer the new kind of retrovirus was made against a background of negative information, which argued against any such major advance. In the first place, we had our experiences with the HTLVs. For some reason, HTLV-1 and HTLV-2 could not, as far as Mika or I knew, be added to and grown in an already immortalized cell line. Although either one of the HTLVs can infect normal fresh blood T-cells and occasionally *induce* them to immortalize, and HTLV can be continually produced from these cell lines, we never transmitted either HTLV to an already established cell line. Second, the French group insisted that LAV could not be grown in an immortalized cell line, and of course could not itself induce cells to immortalize. If our detections of the new kind of retrovirus turned out to be of the same type as LAV, as we now suspected it would, this too argued against success. However, Mika now began to question that assumption seriously. After all, what did he know about the French attempts? Very little. What cell lines had they used? How many attempts had they made? Had they co-cultured with cell lines not immortalized by another virus? Had they used adult T-cells? What protocols had they followed? What exactly had been their results? They had not presented this information in the scientific literature or in oral presentations—only the negative results were mentioned. Perhaps they simply had not tried hard enough.

Increasingly, as I have suggested, it seemed the only explanation for CC was that it had two separate viruses—an HTLV-1 immortalizing the T-cell and a second, possible AIDS retrovirus, struggling to kill the immortalized cell but not quite succeeding. But wasn't this, in essence, an example of a possible AIDS virus growing in a cell line?

The failure of one earlier experiment kept coming back to Mika, reinforcing his feeling that the only way he was ever going to solve the culturing problem was by infecting a cell line. In that experiment, he had attempted to keep growing the co-culture of baby blood cells Phil Markham had obtained from Yale, from an infected mother and child. As he tells it: "When Markham told me about the baby's blood cells, I was very excited. Here was what I was looking for: a sample of blood that was likely to be free of any other virus—no CMV [cytomegalovirus], for example, nothing to confuse the results. Surely the putative AIDS virus would be easier to find in such baby blood. Even though the sample Phil gave me was small, I further reduced it by fractionation, because I wanted only the CD4 lymphocytes, the target of the AIDS virus. I was really hopeful that we would succeed with this one. But no luck. Later, when I factored in the results of

Robert-Guroff's serology and Gelman's molecular study, I knew for sure that this new virus was too cytopathic ever to grow easily in a short-term co-culture. What we needed was a cell line strong enough in its own reproductive capabilities to overcome the cytopathic effects of the virus."

Mika later stated that he had decided that he was going to try to get one or another isolate into a cell line, no matter what it took. If necessary, he would attempt to infect every cell line in the lab with as many isolates of the virus as he could get his hands on, in as many combinations as necessary, until he found a cell line that would continue to grow. It was now our collective view that this kind of success would be absolutely essential if we were to find out whether these various retroviruses, all different from HTLV-1 and HTLV-2, were each of the same kind—that is, closely related to one another—and whether they caused AIDS. It would also be absolutely essential for the development of a life-saving blood test and as a cell-virus permanent system, allowing for the first time the proper tracking of the epidemic.

Mika's own earlier detections of samples that were negative for HTLV-1 and HTLV-2, and positive for reverse transcriptase, had not been maintained in culture long enough to be of much use now. What was available were small amounts of frozen cultures, but whenever virus is frozen and thawed, some fraction of active virus is lost. He did not know that Markham and Salahuddin had made substantial and rapid progress in culturing the primary AIDS-infected cells in human umbilical-cord blood or simply as primary cells.* Back in September, shortly after we had returned from Cold Spring Harbor, the second sample of LAV arrived in our laboratory from the Pasteur Institute for Mika, and that too he put in the freezer where it sat for close to a month.

Betsy recalls that just before Mika left for Switzerland, where he would be picking up new samples from Zairian patients in Basel, he told her that he was again thinking about cell culturing with neoplastic cells in the hope of finding a permissive cell line. He had wanted her to take LAV out of the freezer and, following his protocols, try to put it into five different cell lines. He hadn't looked at LAV since its

*The two groups were not always keeping each other informed, partly due to a bit of rivalry. By January, I put an end to the separate approaches.

arrival. He hadn't even told me it had arrived. But he knew it was there, as free virus particles in supernatant fluid form (we had not received cells that produced LAV), which made it ready for an attempt to infect cells.

Right before he left, on October 20, Mika had cultures set up with his own ongoing isolates and Betsy took care of them as well, in preparation for more experiments. During this time, Betsy began trying to transmit LAV, following his protocols, to five different cell lines. He would not return for almost ten days.

Five different leukemic T4-cell lines were selected for the first transmission because some existed as immortalized cell lines and because T4-cells appeared to be the major target cell of the AIDS virus. One problem was that the vast majority of established human T4-cell lines were infected with and immortalized by HTLV-1 or HTLV-2. They would not be very useful because they would always yield a mixture of the new retrovirus type with an HTLV. Only a few human leukemic T4-cell lines not infected with HTLV-1 or HTLV-2 and not thought to be contaminated with these viruses were available anywhere in the world at that time.

One, HUT 78, was established as a cell line in the late 1970s by Paul Bunn, Adi Gazdar, and their co-workers at the National Cancer Institute and characterized with a small amount of help from investigators in my laboratory. A second, called Ti7.4, was established in culture by Abby Maizel and his colleagues at the M. D. Anderson Hospital Tumor Institute in Houston. Two other lines, called CEM and MOLT-3, were established in England and in the United States, respectively. They were chiefly characterized by Jun Minowada, a Japanese medical scientist then working at Roswell Park Memorial Institute for Cancer Research, Buffalo, New York.* All these cell lines came from patients with malignancies involving T4 lymphocytes. No one knows why these cells permanently grow in tissue culture. Since they are HTLV-1–negative and HTLV-2–negative, the mechanism by which they became immortalized is unknown.

Of the five cell lines Betsy attempted to infect, two died within days. A third was near death by the time Mika returned. But the remaining two showed retrovirus production, though low level, and what she

*Jun Minowada was then and continues to be one of the most generous biomedical scientists I know. Early in our HTLV-1 work, he was critical in establishing the rapport and active collaborations that showed that the virus found in 1981 in Japan was the same as our HTLV-1.

called "ballooning"—referring to the syncytia, the giant, multinucleated cells often characteristic of our earliest culture of AIDS cells and indicative of the presence of a cytopathic virus. Neither cell line was particularly strong, and both also soon died out. But this didn't matter. Neither Betsy nor Mika had expected any of the cell lines to take. On his own, Mika started LAV up once again in the two most promising cell lines. He also called Montagnier to report that the EM report showed viral particles and that "we know how to work with this kind of virus."

Once Mika knew that he could get the virus into cell lines, even temporarily, he would repeat and extend the experiment. Within a week, he would succeed three times.

It has been suggested, by a couple of members of the press and by patent lawyers later hired by the French, both that we had no business asking for LAV and experimenting with it and that we were irresponsible scientists for not immediately recognizing its importance and working with it exclusively. We have also been asked why Mika transmitted LAV first, and not one of the isolates that were coming either from his own earlier work or from other areas of my group, such as Phil Markham's lab. It has also been suggested that the real reason we asked for LAV and Mika told Betsy to try to get it into culture was that we had no isolates of our own to work with. Nothing could be further from the truth.

There was nothing unusual in our request for this isolate; our lab's confirmation of the French findings helped them more than it helped us. We lent credibility to their work, not the other way around. As important, the French would not have sent us LAV if they had any doubt about the propriety of our doing experiments with it. Why did Mika use LAV? Partly, as I have said, because it was there and ready to be transmitted to a cell line without further work. Partly because Mika's stock-in-trade was in doing what Montagnier said could *not* be done—growing the virus in a permanent cell line. And partly because of curiosity about what they really had. But LAV was not sent for collaboration. It was sent in a spirit of cooperation; but, more important, because the French had published results regarding LAV, they were obligated to send it to us upon our request. Early in 1984, however, I directed Mika to stop working on LAV and focus on the isolates from our own laboratory, partly because I wanted him to focus on our own material and partly because of the July discussion with

Montagnier that left uncertain whether we had approval to publish results with LAV.

I gave very little thought to having LAV from the French lab in our lab. Not for a moment did I think that someday it might look bad for us that the French had sent us their virus. Hadn't we sent them cell lines, HTLV reagents, and IL-2, as well as providing technical information and ideas as they began their work? I simply thought that since they had sent it, we'd do some analysis with it. Perhaps it would be a useful reagent for us, as things we sent to them were sometimes useful to them. We had our own direction and our own plan. I wasn't asking Mika every week, What's the latest on LAV? Mika didn't need my permission to work with it. We were not focused on LAV: Mika did not even bother to look at it until a month after receiving it, and I did not know about his culturing it in target cells until late December, two months after the fact.

Major successes were also occurring elsewhere in our group. By early fall 1983, Phil and Zaki had solved one part of the culturing problem—why we were not *consistently* finding evidence of reverse transcriptase activity in blood cells from AIDS patients. In many respects, our past experiences with HTLVs had helped us greatly, but not in this one. HTLVs normally produce steady but low levels of reverse transcriptase after several weeks of culturing; but in the case of the AIDS virus, the RT level spikes. The difference in what you get when you set up the assay properly for HIV from HTLV is like the difference between a mud puddle and Lake Michigan. Unlike the HTLVs—which, having produced virus, maintain extended viral production for a long period—peak production with the AIDS virus occurs during a shorter period, after which the primary cells often die.

When facing a virus of unknown properties, a researcher should try the widest variety of conditions. But at the time it escaped my attention, and we did not do this until Phil and Zaki began their work. This was costly. The virus was more cytopathic than we had anticipated and, by the time most of the samples arrived from our medical collaborators and we started looking for RT activity, we had often missed the retrovirus peak period.

In contrast, when we started getting fresher samples, which enabled us to do earlier testing, Phil and Zaki began seeing extensive retroviral activity and getting their isolates to grow for longer and longer periods. "What we found," recalls Markham, "was that early on—in June, July, August of 1983 [before the Cold Spring Harbor meeting in September]—we had so many samples come in that we were able to see a pattern and with each new insight, we went back and changed things. By October of that year we were able to keep something going and producing the new kind of retrovirus indefinitely, and we kept that culture going into the next year even without transmitting the virus to an immortalized cell line."

In November 1983, after Mika returned, Betsy, working in another section of the lab, also began putting other isolates from individual patients into cell lines. One that she worked with carefully was "RF,"* which became the second well-characterized isolate of the new type of retrovirus isolates. She grew RF in blood cells (not a cell line) for several weeks before it started declining rapidly. At that point, Betsy took the initiative to put it into a previously successful cell line—HUT 78. That too took, and virus production got more and more impressive. She remembers going to Mika and proudly showing him her own culture: "Great, great," she recalls him saying. Later, when Betsy was to send these cells out for other tests, she would refer to them as "Betsy's cells."

"Betsy's cells" were to play an important role in being compared with one of the other early viruses put into a cell line. As a consequence of work performed primarily by George Shaw, Beatrice Hahn, and Flossie Wong-Staal, the lab would be able to report for the first time the discovery that is now a hallmark of the AIDS virus, its genomic heterogeneity—that is, the likelihood of finding great variance among the nucleic acids of different AIDS virus particles, something I will speak more of in chapter 14. At the beginning of November, Mika started another isolate, MOV, into two cell lines, HUT 78 and Ti7.4. Again he was successful.

The name MOV had a different origin than most. Mika often described a cell culture containing virus in his lab notes simply as *V* for virus. When the virus was successfully cultured for any period of time, and the results from an RT assay came back positive, Mika then called his culture *VI,* virus-infected. When Sarang did some further work on this particular VI, he reported that he had detected what he thought

*Many different code names were used for different virus isolates at that time. RF is the initials of a Haitian patient with AIDS whose cells were sent to us by James Hoxie of the University of Pennsylvania.

was reactivity in this isolate to HTLV-2 reagents. Since HTLV-2 was first isolated from a patient named MO, Mika referred to this isolate as MOV. The possibility existed that the cell line was doubly infected with HTLV-2 and the new putative AIDS retrovirus, either because the patient had a double infection or because of laboratory contamination with HTLV-2–infected cells, maintained in our lab at the same time. Nonetheless, MOV was now also in a cell line.

Before we were able to conclude that the new and different retroviruses—that is, more clearly not related to HTLV from patients with AIDS—were all one and the same virus types (before we could settle on a single generic name), naming these isolates was a problem and could easily lead to confusion. One researcher, for example, called all our new retroviruses LAV, assuming only that they were new, different, had some similarity to the Pasteur isolate LAV, had something to do with AIDS, and were probably all similar to one another. The various isolates from our lab studied by him became LAV (MoV), LAV (3B), or LAV (PZ), for example. Not until we had information that several different detections or isolates of this retrovirus were all one and the same virus, and were convinced that this new virus was the cause of AIDS, did we decide on a formal, uniform, *generic* name, the name that had been suggested at the Cold Spring Harbor meeting, HTLV-3.

The initial experiment with LAV and the earlier studies with CC contributed to our learning to solve the problem of transmissibility to cell lines. This was soon reproduced with several other isolates, and it soon became apparent that all the infections did not result in equal virus production. The early cultures of LAV, as well as some of our own isolates, did not give high enough viral production to be of use for a blood test, or even for purposes of viral characterization. This, after all, had been the key stumbling block all along. Mika now decided to tackle this next problem—how to maintain the viral productivity of the infected cell line.

Clearly, Mika had been able to infect cell lines to get virus production. But repeated infections of cell lines was not the final goal; the problem would be solved only when he had a high level of infection, one that would neither die out nor lose productivity.

By definition, a successful infection of a target cell *in vitro* must achieve a sufficiently high ratio of active virus particle to target cells. This ratio is called the multiplicity of infection, or M.O.I. Retroviruses can degrade easily. Many are defective to begin with. Thus it is not uncommon to try to achieve an adequate M.O.I. by concentrating the virus in the solution vis-à-vis the cellular debris. Now Mika decided to take this thinking one step farther.

In his next experiment, he pooled together several samples. Within a week, he saw a strong cytopathic effect (giant cell formation) and later had positive RT readings. He continually reinfected by adding virus back to the cell line or by adding fresh cells as needed. In addition, at one point it seemed that production might be dropping off. So he pooled virus from several additional patients and added them to the cells. Now he had virus from ten patients in the pool.

My colleague Bill Blattner, a clinical epidemiologist for our AIDS task force, remembers that during several of the AIDS task force meetings, Mika and I had made special pleas for more and better samples, "particularly for fresher samples. People were now responding, so much so that there were literally hundreds of samples coming into Gallo's lab, and people were working routinely until midnight to take care of all the culture flasks. At one point, on short notice, by administrative directive Mika had to change laboratories to make room for an administrator's office. In the middle of this key culturing period, he and all his AIDS culture flasks were shuffled from one laboratory to another. Several cultures were damaged and the cells producing virus were lost; others were likely contaminated. The added pressure of the move made a difficult period even more trying and also resulted in the loss of many months' worth of work.

"It was at this time that Mika was given some special samples, all very fresh. One, from a male homosexual, was hand-carried from Dr. Broder's office, probably by Broder himself. [Broder is now the director of NCI.] A second was collected by NCI epidemiologist Jim Goedert. It included tissue from a lymph-node biopsy, bone marrow, and a large quantity of blood, again hand-carried and delivered fresh. Mark Kaplan at North Shore Hospital went out of his way to supply further fresh samples.

"It was from some of these specially acquired samples that Mika grew up his pool of virus, because individually these samples showed the best cytopathic effects and therefore had the most virus. Later I learned that not all the samples were the best to use, but in the end the good ones made the difference."

As it turned out, the virus from only one patient took hold in the pool. To this day, we do not know which patient it was from.

When the pooled cell line was assayed for reverse transcriptase, we finally had what we were looking for—a cytopathic virus, growing continuously and in quantity in a cell line. Betsy remembers the very moment when the RT results came back from Sarang, and Mika saw just how strongly positive they were. According to Betsy, he turned to her and said, "One day we'll tell our grandchildren about this moment."

On December 29, 1983, Sarang immunized the first rabbit with the new retrovirus in order to get the much-needed reagent, specific anti-AIDS virus antibodies. It was the first of many processes that would allow us to confirm that we had indeed found the cause of AIDS. Soon Jörg Schüpbach, a gifted postdoctoral Fellow from Zurich, working with Sarang, would have all the proteins of the infected cells compared with the uninfected cells to get the first ideas about the new proteins encoded by the new virus. Mika would scale up the culture to provide Sarang with enough virus to isolate and characterize the viral antigens to develop an accurate blood test. Betsy already had RF isolated and in culture. Mika was succeeding with still more cultures, and Phil and Zaki with others. Meanwhile, Mika's one attempt at pooling material from different patients did not let us down. It is faithfully replicating and producing virus to this day, as are others we would soon isolate and also succeed in growing in cell lines.

In the euphoria we were all experiencing, none of us could have foreseen that what should have been an unqualified victory would become one of the most challenged and scrutinized discoveries in modern scientific history.

10

Making Progress, Making Sense: The Period of Intense Discovery

Jonathan Mann, the former head of the AIDS Program for the World Health Organization in Geneva, refers to the years from 1983 to 1985 as the "period of intense discovery" for AIDS research. During these two years the cause of AIDS was definitively established as a single virus, and the AIDS virus was fully characterized. Specific reagents to it were developed. Progress was being made at a truly astonishing rate, particularly regarding a number of important discoveries about how the virus creates disease in humans. And our lab developed a workable blood test with an innovative confirmatory test that became the "gold standard" of the clinical serology.

By early 1984, we named the new retrovirus, which we now believed to be the cause of AIDS, HTLV-3, and the specific virus strain that Mika derived from the pooled material and grew in the HUT 78 cell line, HTLV-3B (the *B* simply stood for the fact that the sample was split into two portions, A and B). Mika was aware, however, that earlier analyses had shown that several different human T-cell lines carried the name HUT 78. Which one matched the cells now carrying the virus? Because of this uncertainty, Mika began calling the cell line he was working with HT (for human T-cell leukemia) instead of HUT 78. Reasoning that some variants of this cell line might be more resistant than others to the killing effect of the virus, and to obtain greater certainty that no HTLV-transformed immortalized T-cells were contaminating the cell line, Mika made a number of single-cell clones of HT.* Indeed, some did prove to be more resistant and eventually the H9 clone infected with the pool of virus supernatant fluid was selected for expanded growth. For technical reasons there is no need to detail here, the virus isolated from the pool and grown in H9 and in some other cell clones and cell lines would now be called the 3B strain of HTLV-3.

In November 1983, Popovic carried out our first limited serological test for antibodies to the HTLV-3 retrovirus. He tested for antibodies in a small group of patients who had clinical AIDS, and obtained positive results. As important as this finding was, it was far from definitive because the numbers tested were few, and, more important, the test he used employed the infected cells as his target for antibodies present in sera of AIDS patients, similar to the test Essex had used earlier with HTLV-1. It was a convenient method, but such cells could conceivably harbor other viruses or microbes, and since blood often contains a variety of antibodies as a consequence of past or present other infections, the results might not be very specific. This was particularly true of AIDS patients, who were exposed to a great many infections.

Conclusive serological testing, in our view, required finer, more specific assays based on using purified virus particles of proteins obtained from the virus instead of whole cells infected with virus. It would also be critical at this early stage to develop reagents specific to the proteins of the particular virus being tested for, so that the reverse kind of testing could also be done—that is, testing for a specific viral protein in blood cells or in viruses obtained from a patient, but testing not with patient sera but with a well-defined antibody, one in which the specificity could be precisely defined. This would be our next step. One of the most useful reagents is called *hyperimmune* sera. Animals (rabbits or goats are usually used) are repeatedly inoculated with a harmless part of the viral proteins of the target virus. After each inoculation, the animal is bled. In time, with repeated inoculations, the sera of such animals will show a very high

*When the word *clone* is applied to cell lines, it means that a single cell has been isolated from the mass population of variants present in a cell line. The single cell is then cultured and becomes a cell line itself.

level of antibodies specific to one or more of the proteins of the target virus.

With hyperimmune sera, much more definitive findings were possible. For example, we could now clearly demonstrate that the biologically active agent in different cultures was one and the same type of virus. Conversely, because we now could produce large amounts of well-characterized viruses in a single type of cell, we were ready with sensitive and reliable techniques for epidemiological blood testing for antibodies to this virus in sera. So, using virus itself in the enzyme-linked immune absorbent assay (ELISA) test,* Sarang could now conduct wide-scale serological testing of the general population for the presence of antibodies to the AIDS virus. Our first test group would include people from each of the three target groups: those with AIDS, those in high-risk categories, and a control population.

We quickly determined that even with a quality virus and a stable single type of cell-producing virus (therefore less variable background "noise"), the ELISA could sometimes give false positive or false negative results. A confirmatory test was essential.

The solution came from the innovative work of Jörg Schüpbach, the postdoc in our lab from Zurich. Working with Sarang, he got the idea to make clinical use of a test he had used in the laboratory to tell us what the major proteins of the AIDS virus were, how big they were, and which ones reacted with antibodies in the serum of infected people. He thought that this test—called the Western Blot†—might serve as a backup to the ELISA. Its appeal lay in the fact that it allowed

*This assay, called the ELISA test, was developed by B. van Weemen and A. H. W. M. Schuurs of the Organon Company in Holland. It is based on the following principles: (1) The proteins of the test virus (here it would be the AIDS virus) are obtained from a purified virus sample and are coated on the wells of plates. (2) Test serum samples containing antibodies to the virus are then added to the plates. During incubation, the viral proteins on the plates form complexes with those antibodies in the test sera, which react with them. (3) Following incubation, the sample is aspirated and the wells washed. Antibodies made in an animal against human antibodies (antibodies being proteins, when foreign, can also induce antibodies against themselves) conjugate with an enzyme, which binds to antigen-antibody complex during a second incubation. (4) Following further wash, an appropriate color forming substrate is added and incubated. The color yield is measured using appropriate instrumentation. The intensity of the color is proportional to the amount of antibodies in the test serum.

†The principles of this procedure are: (1) Separation of proteins of the virus by a procedure that takes advantage of differences in sizes of proteins. This is done in a gelatinous substance. Subsequently, the separated protein bands are transferred to a paperlike sheet. (2) Incubation of strips cut from the nitrocellulose sheet with the diluted test sera, allowing complex formation between specific viral protein bands and antibodies in the serum. (3) Incubation of the strips containing the immune complexes with appropriately tagged antibodies to human antibodies. (4) Visualization of the antibody reaction. If the tag is radioactive, the reacted bands are seen upon exposure to and development of X-ray film.

the clinician or blood bank worker to scan blood for antibodies to a variety of proteins by their size. These proteins include parts of the envelope and core of the AIDS virus.

Though ELISA was reliable and could be calibrated to give very few false negatives, it did produce under these conditions a number of false positives. In situations where the patient had antibodies that cross-reacted even with one protein of the AIDS virus, the ELISA test might give a positive finding. Because it is extremely unlikely that in the absence of infection with the AIDS virus, positives would be produced to the range of structural proteins specific to the AIDS virus scanned by the Western Blot test, it produces many fewer false positives.

Done correctly, this combination of tests would be as good a screening device as we could hope for—both sensitive and specific. Our own laboratory trials were performed with coded samples, so that we did not know the diagnosis in advance. The first results we obtained with over 1,000 serum specimens showed that between 88 and 100 percent of AIDS patients' sera were positive (our results varied in different studies); a significant percentage of each of the then known risk groups was positive (this would vary according to the exact region, exact risk group, and date of serum sample collection); but only about 0.1 percent of healthy donors from so-called nonrisk groups were positive.

In a later collaborative study with investigators at the Centers for Disease Control, we found that 100 percent (twenty-eight of twenty-eight) AIDS patients who had earlier received blood transfusions as their only risk factor for the development of AIDS were antibody-positive. Even more important, in every one of these twenty-eight cases, at least one of the blood donors was antibody-positive, and this was during a time when only 0.1 percent or less of the U.S. population was infected. Moreover, many of these blood donors had subsequently developed AIDS.

In other studies of blood from various parts of the world, Sarang and Schüpbach found serum antibodies in blood from regions where AIDS was present, but little or none in blood from places where AIDS had not yet occurred. These results and these assays became the basis for the blood test for the AIDS virus.

During the same period, I made plans for more extensive efforts at virus detection. Our previous experience with HTLV-1 had shown that the presence of serum antibodies to the virus always indicated infection.* Still, I did not believe that serum antibody results, no matter how correlative the positives to the groups within which we expected to find the virus, would be sufficiently persuasive to the main critics in the scientific community with an epidemic disease like AIDS. I felt it was important to demonstrate that the same strong correlation of serum antibodies to virus isolation and hence to proof of infection we had found with HTLV-1 was true also with the new retrovirus, the presumed AIDS virus. I knew that if this retrovirus was the cause of AIDS—as we were now convinced—we would need to convince the academic community as totally, as widely, and as quickly as possible. Not to do so would mean much wasted time, money, and effort—and, obviously, loss of life. Further, at the time of this work, multiple theories for the cause of AIDS still abounded. From this point on, we worked with a feeling of great pressure and speed.

Much earlier, I had outlined plans for Zaki Salahuddin and Phillip Markham to survey tissues of AIDS patients, risk groups, and normal controls for the presence of virus. We were already confident that blood cells from AIDS patients positive for RT and negative for HTLV proteins (having tested them with specific antibodies made earlier against the HTLVs) were likely producing the new retrovirus. We would be even more confident when in some instances we were also able to show that the virus exhibited a cytopathic effect and in some cases that the cells contained proteins that reacted with our rabbit hyperimmune sera.

In a few cases, the presence of the virus was further verified by electron microscopy carried out by others in collaboration with us, but electron microscopy was not routinely used.† None of the tests produced evidence of the virus in any of 115 normal blood cells cultured from healthy volunteer donors.

We concluded that this new retrovirus was the cause of AIDS; that our blood test was useful in preventing further medical spread of the virus; that a laboratory system was now available for testing the effec-

*This is an important point. In most other kinds of infection, an antibody response need not mean persistent infection; often it means the opposite: the individual was once exposed to the microbe, perhaps even heavily infected, has resisted it, and the antibodies are an indication of this continued protection. In fact, the failure to appreciate this difference with retroviral infection led to some unfortunate early disclaimers by some health care workers, who stated that the positive results of the AIDS antibody test did not necessarily mean infection, and that no one really knew what the result meant. We did know: it meant the patient was infected persistently.

†When it was used, electron microscopy was usually insufficiently sensitive as a screening method.

tiveness of various drugs against the virus, for eventual use by AIDS patients; that the spread of the causative agent and hence the epidemic itself could now be monitored for the first time; and that there was now sufficient virus to think about a vaccine program.

We named the virus the third human T-lymphotropic retrovirus, or HTLV-3. We called the continuously cultured sample in H9-cloned T4 leukemia cells HTLV-3B and the sample from a single patient Betsy had cultured, HTLV-3RF.* Other examples of isolates that would soon also become stock prototype AIDS viruses from our laboratory included HTLV-3MN, HTLV-3SC, and HTLV-3BAL. Many others (over fifty) were cultured only for short periods either because of technical difficulty or because growing many more viruses of the same kind would be redundant and not serve a useful major objective while stressing our resources to their limit. We strongly suspected that all of these isolates were of the same general type of virus as LAV, the BRU isolate. I held off doing the comparison for a short time to allow a collaborative project with the French group after our own work was written. Although at the time I thought it was the proper way to recognize the French contribution, this turned out to be a mistake.

Looking back on our work over those few years, I am sure beyond any doubt that other than our early and, one might say, baseline contributions to the field—such as the idea to look for a retrovirus in AIDS and some technical contributions, such as the growth of T-cells in IL-2—of our key contributions, the most important new technical advance was the ability to produce the virus continuously in a defined, immortalized cell line. Relying on umbilical-cord blood T-cells and IL-2 (our older methods for starting HTLV cultures and employed by the French group with their LAV isolate, modified by frequent replenishment of the normal cord blood T-cells) could never solve the problem. I think our second most critical contribution was the breadth of the approaches we took so as to leave a critical scientific community and (some segments of) an increasingly skeptical society no choice but to conclude that the new retrovirus was the cause of AIDS.

In early 1984, before writing our papers for publication, I informed the NCI director, Vincent DeVita, that we had succeeded in putting the AIDS virus in such a cell line and that we now were sure we knew

*When it is obvious that the AIDS retrovirus is being referred to, the strain abbreviation alone is often used, that is, 3B, RF, BRU, and so on.

the cause of AIDS. I reviewed the evidence with him. But I asked him to keep this information quiet until our papers were published in the scientific press.* He agreed but asked that I too keep the information quiet at this stage because otherwise all sorts of rumors would "float." He also asked that an exception be made to inform Assistant Secretary of Health Edward Brandt. I also told DeVita in detail about the work of the Pasteur group.

March 1984 was a time to consolidate our results and to put in final form the five papers (four for the U.S. journal *Science* and one for the British clinical journal *Lancet*) we would be submitting to announce our findings. Just before this, in February 1984, an AIDS meeting was about to take place in Park City, Utah. As a chairman of the meeting, I was scheduled to give an introduction. Representing the Pasteur Institute was my friend Chermann, who spoke about the Pasteur's latest findings of additional detections of "LAV-like retroviruses." No evidence was presented that they had made progress in propagating LAV (BRU) in a cell line, and their testing of AIDS patients' sera for antibodies to the one they continued to culture (BRU) in repetitively replenished blood T-cells still did not show firm linkage to AIDS, though it was increasing. Nonetheless, his new detections were in risk groups. Clearly, he had made progress.

I used my time at the podium to give a brief overview of the nature of animal retroviruses and the HTLVs and to describe what had been published on AIDS. I also speculated about our work, and theorized that the cause of AIDS would likely be a retrovirus differing from HTLV-1 and HTLV-2 chiefly in the right half of the virus's genetic information, namely, in the region of the regulatory genes and in the envelope. This was roughly the same notion I had spelled out to DeVita in an August 1983 official memorandum.† I also talked for a

*Most of the papers were to be submitted to the journal *Science,* and this journal, like some others, has a policy that if information appears first in the press, the scientific papers may be rejected even after they have been accepted. I knew if DeVita were to mention these results widely, they would be in the press. I also knew they would likely be misrepresented in the lay press due to misunderstanding of the results. As important, other scientists would be disturbed that these findings had been released to the public before they had had a chance to examine our data. I also wished first to have a chance to discuss the major conclusions of our results with the Paris group. Finally, an announcement outside the peer-reviewed literature would not have sped up the blood test; until other scientists confirmed our findings, no responsible official would act on them.

†When the AIDS retroviruses were analyzed, the major differences from the HTLVs were indeed in this region and in those genes, but as we will soon see, the differences were more profound than I ever dreamed.

minute or two about the general implications of Chermann's presentation. But I did not believe I should yet reveal our findings.

There was one person at Park City who did, I felt, need to know that I now believed we had sufficient data to say we knew the cause of AIDS. That was James Curran of the Centers for Disease Control. I asked him to send us coded samples of patient sera from patients with AIDS, from those with other diseases, and from healthy controls. A few weeks later he did, and a few weeks after that I asked to meet with him. The code was broken, and over lunch in Bethesda one March afternoon, Jim Curran found out that we could reliably identify infection by the AIDS virus.

Shortly afterward I headed for a meeting in San Francisco. While there, I received a concerned call from the NCI director's office. Vince DeVita wanted reaffirmation of our findings that a retrovirus was the cause of AIDS. An NIH team headed by Ken Sell, now at Emory University in Atlanta, had publicly announced that the cause of AIDS was very likely a fungus producing an immunosuppressive substance. The fungus was of a new kind that released *cyclosporine,* the one agent medical science knew of that could selectively harm or even kill $CD4^{+}$ T helper lymphocytes! I had immediate palpitations, but soon realized that the team had to be wrong. Didn't Einstein say something about Mother Nature not being cruel but not being so simple, either? Cyclosporine was the one T-cell inhibitor known in biology. It seemed too good to be true. More decisively, we knew from Curran and others that some hemophiliacs got AIDS from infusions consisting solely of Factor VIII, the clotting factor absent from their own blood. This blood-derived factor is filtered in a manner that would exclude fungi but it would not exclude a virus.

The fungus story was, of course, a mistake.* Nonetheless, this episode in and of itself conclusively demonstrates how far away the medical science community still was in the early period of 1984 from accepting a retrovirus or anything else as the cause of AIDS, as do the reports from Poiesz and others at this late period linking HTLV-1 itself to AIDS (discussed earlier) and other persistent, wide-ranging theories.

Our papers were all submitted for publication. In our first paper,

*Apparently, the same new strain of fungus contaminated their culture flasks. The cyclosporine presence was deduced from a spectrum analysis of the culture fluid, which I was told came about from some chemical contamination of the culture.

we reported the development of H9, the clone of the HUT 78 cell line, and discussed its continuous high-level production of virus. In our second paper, we announced that we had detections of this new virus in 48 subjects: in 18 of 21 patients with conditions that were increasingly being defined by clinicians as *pre-AIDS,* especially when found in the defined groups now seen to be at increased risk for getting AIDS; in 26 of 72 patients with AIDS; in 3 of 4 clinically normal mothers of children with AIDS; and in 1 of 22 healthy homosexual subjects. We reported finding no evidence of the virus in the blood of 115 healthy heterosexual subjects.

We also discussed why we thought the virus should still be classified as an HTLV, pointing out that it showed an affinity for T-cells, that its reverse transcriptase was similar to that of the HTLVs, and that there were similarities in the structure of the AIDS virus and the HTLVs that were visible in electron micrographs. We also named those areas in which the virus appeared to be distinct, including the fact that it differed from the HTLVs in its biological effects on the cell and in the immune response it triggered in target cells.

Our third and fourth articles in *Science* characterized the viral antigens—those proteins in the envelope and core of the virus that generally provoke antibody response. We identified five proteins that reacted with antibodies: p65, p60, p55, p41, and p24. The fourth paper specifically showed that 88 percent of AIDS patients and 79 percent of pre-AIDS patients showed these antibodies; one in 186 control samples also reacted. We also identified one particular viral protein, p41, as the one that most strongly elicited an antibody reaction. These *Science* papers were all published in May 1984. The last paper, in *Lancet* (June 1984), reported that 100 percent of AIDS patients showed antibody reaction to this one protein.* These tests of patients' sera were usually done with samples sent to us coded (a "blind" study). Until the code was broken, we had no idea which were AIDS, which healthy controls, and so on.

We should have then just sat back and waited for the publication

*In these studies we had much cooperation and help from our clinical-scientist collaborators, among them Bob Redfield of Walter Reed, Jerry Groopman of the New England Deaconess, Bart Haynes and Tom Palker of Duke University, Gil White of North Carolina, Bijan Safai of Sloan-Kettering, Mark Kaplan of Cornell, James Oleske of the New Jersey Medical Center, Sam Broder and Gene Shearer of NCI, and James Hoxie of the University of Pennsylvania deserve special mention. Because our sera data were already quite extensive, I elected not to use the most recent results obtained with the sera sent to us by Jim Curran.

of these findings. Instead, there followed one of the most trying times of my career, a three-month period that stretched into three years of extreme stress. Ironically, this period was also among the most productive of my life.

By March 1984, DeVita, Brandt, Curran, and a half-dozen people in my laboratory knew some or all of the results. In that month I was also given an unexpected birthday party in the laboratory, which lasted several hours. The entire group now knew about the developments, and people felt confident, proud, and happy. We eagerly awaited the practical applications sure to follow our results, such as working on the development of a vaccine and an effective drug therapy.

I now decided personally to inform the group at the Pasteur Institute. In the first days of April, after participating in a scientific symposium in Zurich, I headed for Paris. I gave a seminar there, reviewing the leukemia story of HTLV-1 and HTLV-2, concluding with a brief discussion of the new virus I called HTLV-3. I stated that I believed our several isolates of HTLV-3 would be the same as LAV, that these were the clear-cut cause of AIDS, that we had a blood test, and that soon direct comparisons would be made. There were, however, some major differences. The key protein in the virus for the blood test by our studies was one we called P41. Montagnier did not show evidence for this protein and stated that antibodies were not made against such a LAV protein. Also, their testing for serum antibodies to LAV was still much lower than ours. I thought, however, these differences might be due to their inability to obtain sufficient viral proteins because they could not grow LAV in a cell line.

Montagnier and Chermann were elated. As Chermann has often stated, up to that moment he had not been able to convince even members of their own institute that they were on the right track with LAV, and as I pointed out, Montagnier himself believed and speculated at the end of 1983 that the cause of AIDS probably involved many different kinds of retroviruses rather than LAV being a specific cause. Although they were making progress, they had been clearly unable to do large-scale production and fine characterization of their isolates. They had recently begun a fruitful collaboration with the CDC to improve their antibody results, which ultimately led them to similar results with their virus to those we had already reached with ours in terms of the blood testing—but still not with the same precision in cases of AIDS.

I informed the French group that my intention was to publish our results, that they were abundant and would be likely to draw considerable attention. I suggested that we make a joint statement after publication and direct comparison of the various viruses on both sides of the Atlantic. This was well received by all, and the mood was very high.

In a few days I was scheduled to attend a meeting in Cremona, Italy, devoted to tumor viruses. I was to be accompanied by James Wyngaarden, the NIH director, but before leaving I agreed to do an interview with Martin Redfearn, a British freelancer then doing a program on cancer genes (oncogenes) for the BBC to be broadcast in June. He interviewed me on this subject but then, at the end, asked whether there was anything new on AIDS. I said yes because he told me that whatever I gave him would be for broadcast specifically in June. This would not conflict with our scheduled publications.

In the euphoria of that time, when nothing seemed capable of going wrong, I made a mistake: I talked with Redfearn, gave him an idea about where we were, and told him he could take materials from our office library. He took copies of the papers that were in press for *Science.*

The meeting in Cremona began with interesting science and many discussions with old friends. Wyngaarden, who by now knew the details of our work, accompanied me and gave an address on NIH support of virology and its international role. The civility and the quiet meditative mood induced by the world's most beautiful piazza and by the music played for us on an original violin of one of Cremona's hometown heroes, Antonio Stradivari, were quickly lost in the din of the bombarding U.S. telephone calls to Wyngaarden and myself.

New Scientist, we were told, was about to print an article with extensive data about our findings—culled, of course, from our forthcoming *Science* articles. Apparently its source was the freelancer Martin Redfearn. The editor of *New Scientist,* Fred Pearce, according to an article written by William A. Check for *The Newsletter of the National Association of Science Writers* (July 1984), made the decision to run the story on the basis of "articles that began to appear in the international press."* Further, it now appeared that some British newspapers also knew the outline of the story. Word reached Washington health officials. Despite attempts by British scientists—such as the virologists William Jarrett and Tony Epstein (discoverer of Epstein-Barr virus, EBV) and Walter Bodmer, a renowned geneticist and director of the Imperial Cancer Research Fund Laboratories in London—to convince people at *New Scientist* of the inappropriateness of their plans, we were told that the story would immediately go forward. It did.

Secretary of Health Margaret Heckler had not yet been adequately informed, and *Science* had an embargo on any public information until roughly one week before the publication date. (There was still about one month to go.) Fortunately, *Science* understood the pickle we were in and the papers remained in press. With the pressure of endless telephone calls from journalists with "inside" information and the calls from co-workers and health officials, Wyngaarden and I were obliged to abandon the Cremona conference and return home.

The decision to hold a press conference to announce our discovery of the AIDS virus was made by Secretary Heckler. I was anxious, not only about how my role in the leaked material would be seen by my bosses but also because of my discussion and agreement with Chermann and Montagnier about a joint statement, roughly in June—an appropriate amount of time after our publications and after proper comparisons could be made in both our laboratories of each of the several isolates we had.† No announcement or press conference had been planned to coincide with the publications in *Science.* But if it were true that the cause was known and a blood test developed, and if it were appearing in British newspapers even if they had only meager information, Secretary Heckler probably had no other choice.

The meeting on April 23 outside Heckler's office and within the press conference room itself reminded me of Federico Fellini's *Satyricon.* I had never been to a press conference before in which research work from my laboratory was being announced. Maybe this heightened its peculiarities for me.

*I do not know whether these articles actually existed. Pearce had a point, though, because a few weeks earlier, Bob Redfield, a friend and clinical collaborator at Walter Reed, had discussed some of the results with people at Walter Reed. Later, he told me someone innocently but excitedly and openly told others at a Washington cocktail party about it. Apparently, the story did circulate and reached at least some newspeople.

†In fact, Montagnier and I did hold a joint press conference in June 1984, during a meeting in Denver. I stated that our several isolates were almost certainly the same virus type as his LAV/BRU.

It began with Heckler's assistant, Mack Haddow, privately scolding DeVita, Wyngaarden, and me, claiming we had ruined the announcement. Not knowing what he was talking about, we asked for information. He provided us with a copy of the Sunday edition of the *New York Times.* It was now Monday, April 23rd—a day later.

On the front page of the Sunday edition was a headline by the *Times* medical science writer and former CDC staff member Dr. Larry Altman, stating that the Pasteur Institute had found the cause of AIDS, although no Pasteur Institute scientist had made such a claim and I knew they had nothing going into publication. The article made no mention of any real new data (no data were published by the Pasteur Institute until a few months after our publications, and what was then published, although important, did not provide a clear linkage to the disease or evidence of a virus in a cell line). The *Times* did not discuss our work at all.

Reading the *Times* article, I could not imagine how Haddow could have arrived at the conclusion that the NIH had preempted Heckler by making a statement solely on behalf of someone else, namely, the Pasteur Institute! (Moreover, the source of the information the *Times* had used became clear to me from a close reading of the article.) Since Wyngaarden and DeVita were political appointees but I was not, I responded to Haddow, asking whether he knew the difference between NIH and the Pasteur Institute. In time, things calmed down a bit, and the press conference began.

Entering the room we faced a sea of cameras and reporters. Nobody appeared to be in a very good mood. James Mason, head of CDC, spoke, and I think Wyngaarden did also. Here are some excerpts from Heckler's comments: ". . . the probable cause of AIDS has been found. . . . a new process has been developed to mass-produce this virus. . . . we now have a blood test for AIDS which we hope can be widely available within about six months. We have applied for the patent on this process today. . . . we can now identify AIDS victims with essentially 100 percent certainty. . . . the new process will enable us to develop a vaccine to prevent AIDS. We hope to have such a vaccine ready for testing in about two years. . . . And as is so often the case in scientific pursuit, other discoveries have occurred in different laboratories—even in different parts of the world—which will ultimately contribute to the goal we all seek: the conquest of AIDS. I especially want to cite the efforts of the Pasteur Institute in France, which has in part been working in collaboration with the National Cancer Institute. They have previously identified a virus which they have linked to AIDS patients, and within the next few weeks we will know with certainty whether that virus is the same one identified through the NCI's work. We believe it will prove to be the same."

I spoke too. Essentially, I confirmed what Secretary Heckler had said—that in my mind the cause of AIDS was unequivocally a new retrovirus, which we called HTLV-3; that this virus was very likely one and the same as the one found by a group of scientists at the Pasteur Institute. I named those investigators in our group responsible for the finding. I also confirmed that a reliable blood test that could quickly save lives had been developed.

Heckler has been repeatedly criticized for predicting when we would have a vaccine, and mocked for her ignorance of how long such things take. But it was not her fault. We (especially I) were her advisers on this, and no one made clear the difference between what was theoretically doable and what was probable. With a causative virus in permanent cell-line culture available to many groups for mass production, a crash program for a vaccine could have been feasible in a few years. No one could have predicted the particular challenges the AIDS virus would present, especially the fact that only a few animal models would be available for these kinds of studies—and those very difficult to work with—and that there would turn out to be considerable variation in the virus from one isolate to the next.

Questions from the press had clearly been influenced by the *Times* article. Reporters directly asked about our relationship with the French. We tried to explain the contributions of both labs and our plans for further collaboration. Massive publicity—some reasonable and some not—followed the *Times* article and the press conference, and there were still four weeks to go before the publication of our results. Much of the press coverage that followed our announcement failed to mention the crediting of the Pasteur Institute. Others in the press, quoting their own erroneous colleagues, even alleged that the Pasteur group was not mentioned at all. Oddly, few seemed to be bothered by our having been totally left out of the *Times* article.

With all the fanfare, a few of the Pasteur scientists were justifiably disturbed. Where were our data? (It would be some weeks before our papers would be published.) Did we really have results to draw the conclusions we had? Did we really have a workable blood test? In addition, they were feeling the natural pain of having enjoyed only minimum acceptance of their earlier belief in LAV. It would get worse before it got better. I would be accused of stealing credit from Montagnier, of claiming to be the sole discoverer of the AIDS virus—as though I thought no one else in the world had participated in the work and, even more ludicrous, as though I have the power to deny credit to other scientists for their own work. Most ludicrous of all were publications by John Crewdson of the *Chicago Tribune* that there was a lost year, as though I alone could and somehow did prevent the Pasteur group from providing the data that could convince their own Pasteur Institute colleagues—let alone the whole scientific community—that they had found the cause of AIDS. On both sides of the Atlantic, scientists made the mistake of trying to clear up questions with newspersons who had no interest in resolving them and calming the story down. Each side allowed itself to be churned up by questions from the press, and often one side or the other responded to a provocation in a way that was less than fair to the other side. It was transatlantic volleyball with too many hard serves.

For a period of time, we could go back to the laboratory and continue what we wanted to do and had proved we could do—learn about the AIDS virus. For the next year, this is what we did. There was much to learn and our discoveries would be both scientifically important and intellectually fascinating. But if we thought we had left behind the world of politics and press, we would soon learn that we had only been given a breather.

With our cell-culturing problems solved, many of the most important findings of the next twelve months would come out of the work of a number of postdoctoral Fellows broadly guided by myself and, in some of the molecular biology work, both broadly and daily by Flossie Wong-Staal. In the meantime it was decided that once our papers were published in *Science,* Sarang should bring our cell-line culture to Montagnier for his use for scientific studies and for comparisons to LAV. In early May 1984, Sarang went to Paris. Our plan was to perform the same comparisons here against the LAV we had received earlier. Sarang left our H9 cell line producing the AIDS retrovirus, HTLV-3B, with Montagnier—and began the experiments.

Once the viral genome became available in purified form, the next step was to clone it—that is, to grow it in certain bacteria that would make multiple copies of the genome. By late 1984, this had been done. Now the genome of the virus could be fully characterized and could also be given out to other investigators to study. Rapid advances in determining the genes of several molecularly cloned isolates and the function of some of the genes soon followed. This was chiefly due to Wong-Staal, Sasha Arya, and postdoctorals such as Lee Ratner, Beatrice Hahn, George Shaw, Mandy Fisher, and Bruno Starcich in our group (Ratner and Shaw are Americans; Hahn is from Munich; Fisher from Birmingham, England; and Starcich from Parma, Italy). Some of this work was done in collaboration with William Haseltine at Harvard and Steve Petteway of Du Pont. Independent work by Haseltine and his co-workers Craig Rosen and Joe Sodrowski led to still more insights into the molecular makeup of the AIDS virus. (In the next chapter I will relate how the AIDS virus got a new name, HIV, for human immunodeficiency virus, but I will refer to the virus by this new name from here on.)

Wong-Staal remembers vividly the excitement in her group at this time: "Working with this virus is like putting your hand in a treasure chest. Every time you put your hand in you pull out a gem. It was not like any animal retrovirus and, we would soon learn, in several respects it was not like the HTLVs either."

Over the course of the next few months, our group would indeed discover a treasure chest of fascinating details about this new virus. First we would demonstrate the heterogeneity of HIV; that is, that there is significant variation in the genome of the AIDS virus from isolate to isolate. This is far less so with the HTLVs. Other experiments by Shaw led to the unexpected discovery of genetic sequences of the AIDS virus in the brain. This changed the whole picture of how the thinking disorders known to occur in AIDS patients actually develop. Prior to this finding, most clinicians believed that the abnormal brain function associated with some AIDS cases was due to a variety of causes, including *toxoplasmosis,* a parasitic opportunistic infection. Now we suspected that a direct pathogenic effect of the AIDS virus on brain cells might be involved.

Meanwhile, Sue Gartner, Popovic, and, independently, Markham and others in our group discovered that the virus infects not only the T4 lymphocyte but also macrophages.* Considering the demonstra-

*Remember, macrophages are cells that, like lymphocytes and granulocytes and red blood cells, come from a primitive cell of the bone marrow. Their lineage is more connected with the granulocyte, forming the lineage we call myeloid. Macrophages are found in blood but because they have the capability of moving in between cell junctions, they get to tissue sites as well, often

tion of viral sequences in the brain, we determined that the cell known as the microglial cell was a target.* It was already believed that microglial cells were closely related to macrophages. Things were beginning to make sense.

In a very important discovery, Robin Weiss in London, with his co-workers Gus Dagleish and Peter Beverly, as well as David Klatzmann working with Montagnier in Paris, showed that the receptor for the virus was a molecule present on the surface of T4 helper lymphocytes and known as CD4. We and others found that macrophage and microglial cells had the same molecule but in lesser amounts on their surface. Another corner of the picture in the jigsaw puzzle began to emerge.

About this time (June–July 1984) Wong-Staal compared the genetic material of LAV with our isolates. In Paris and in our own lab, the proteins were compared. We concluded that LAV and our isolates, generically called HTLV-3, were of the same virus type. The plan was to publish two papers on this comparison, but when it came time to do so Montagnier changed his mind. During this period both of our groups were close to finishing up fine analyses of the genetic structures of these viruses, including determining the exact sequence of nucleotides of our viruses. He believed that the availability of the genetic sequence made any other kind of comparison by us redundant and far less precise: a scientist could just look at the published sequence and see the close relatedness of the viruses; the order of the genetic information is after all the ultimate comparison. Since this logic seemed reasonable, we agreed. Later I would regret this decision.

Our fine sequencing studies were completed in late 1984. Credit for this work goes to Wong-Staal's group, working in collaboration with Bill Haseltine and Joe Sodrowski at Harvard and others.† We

---

responding to chemical attractants which, for instance, are released by some cells after local infections. Macrophages play a pivotal role in our defense mechanisms against microbes and other small foreign bodies. One of their major functions is to take foreign protein, process it (by cutting them up into smaller pieces), and present certain specific pieces on their surfaces to be recognized by and induce stimulation of other cells important to the immune responses, such as the T4 lymphocyte.

*The best-known and functionally central cells of the brain are called neurones, but several other structurally and functionally distinct cells are also identifiable. Most of these form the bulk of the so-called white matter of the brain. Their precise functions are not completely understood. Among them are *microgial* cells—relatively small cells, believed to be macrophages, that enter the brain from the blood circulation. How often this occurs is not known, but it intensifies after injury.

†In our group, Lee Ratner played the major role. Other important contributions were made by

learned that there were 9,749 nucleotides in the AIDS virus genome—to our surprise, it was significantly larger than the HTLVs. We also learned that although it too had a very interesting gene that seemed to parallel the *tax* gene of the HTLVs, it was located in a different position on the AIDS virus (later it was called the *tat* gene). There also proved to be regions of the HIV RNA that suggested that the HIV contained several more genes than the HTLVs. In time we learned it did. Two of them (soon called *tat* and *rev*) were like the extraregulatory genes (*tax* and *rex*) of the HTLVs. Four other extra genes were not like anything in the HTLVs and were totally unexpected. (Later, we would make mutants of the virus—altering certain pieces of its genome, generally one or another gene. By this means we and others were able to show that some but not all of the extra genes of HIV were required for the virus to replicate.)

As I have mentioned, we also proved the virus was relatively unstable; we learned that there was considerable variation in the viral genome when comparing one isolate to another. We would later learn that this instability carried over to the same isolate over a period of time—that is, the genome of a single virus changed even after it infected a cell, suggesting to us just how difficult the production of a vaccine might be. When all our sequencing work was done, we realized we could not continue to characterize the virus as a member of the HTLV family of viruses, though I continued to think (and still do now) that the name HTLV-3 was logical. After all, many human papilloma (wart) viruses have little of their genetic information in common yet are still named the same with different numbers, and various human hepatitis viruses are named in alphabetical order, even though they do not even belong to the same virus class, a far more extreme difference than HTLVs and HIVs. On the other hand, there were, no doubt, many significant differences, and the Pasteur group had first argued this point. Now it was scientifically demonstrated.

Last, and most unsettling, we discovered that one of our own HTLV-3B isolates was much closer to LAV than was typical of our other isolates. George Shaw and Beatrice Hahn, two postdocs with us I have mentioned earlier, had been the first to show, well before the actual sequencing data had come out, that there was much genetic

---

Steve Petteway and his colleagues (then at Du Pont) and by Takis Papas (NCI) and Nancy Chang (then at CentoCor, Inc.).

heterogeneity in the virus from isolate to isolate. They had done their work comparing two of our isolates, HTLV-3B and RF, and then a couple of others, then some fresh specimens from postmortem tissue. Practically all were genetically different from one another. Yet LAV and our 3B isolates were distinctly close to each other.

In June 1984, I called Montagnier and told him of this last finding and of how odd I thought it. He seemed rather indifferent, however, remarking that "they *should* be very close—they were all the same virus, after all." I said yes, but also told him that we'd obtained data showing variations between different isolates, data that would be published in late 1984. He commented that maybe there were different groupings. I realized that one possibility was that LAV/BRU and HTLV-3B were actually the same isolate, maybe an accidental contamination occurring in his or my lab.* I thought his attitude then was appropriate, however, and who would care? After all, there was no doubt that early on he had his own isolate because I knew we had received a bona fide novel retrovirus in 1983 from him, and I knew we had isolates other than the 3B strain. But this assumption was a mistake, one that would cause us many political headaches and much loss of my scientific time. It has also fueled a few writers to this day, particularly one obsessive reporter, the same Mr. Crewdson.

Meanwhile we continued in some particularly important epidemiological studies in collaboration with many investigators. Some of these simply looked at HIV infection in different groups and countries. Other studies in which we participated but which were led by Robert Redfield of Walter Reed showed heterosexual transmission of the virus for the first time. With Daniel Zagury, I reported the finding of virus in the T4-cells or macrophages present in semen. Since such cells are unusual except in males with some forms of venereal disease, we suggested that certain venereal diseases might be a co-factor in transmitting the AIDS virus.

We discovered, and also published in our earliest papers that we

*Stanley Weiss, an epidemiologist now working in New Jersey, was at NCI at that time and happened to have been near my office when I made this call. My concern, he recalls, was more for the likelihood of a contamination in Montagnier's lab, and he confirms the relaxed attitude I have described. Later I had reasons to think that if the two viruses were so closely related that they might have been mixed up (accidental contamination), the mix-up would likely have occurred in my laboratory. In any case, because we had many other isolates and had made several other key scientific advances, not to mention the enormous scientific-medical problems that still lay ahead of us, my co-workers and I did not think this likelihood terribly important.

found, the virus in plasma as free virus particles in a patient with AIDS. This showed that viremia was present, something we had never seen with the HTLVs. Salahuddin and I, in collaboration with Jerry Groopman at the New England Deaconess Hospital in Boston, reported finding virus in very small amounts in saliva. We stressed that this did not change the conclusion drawn from clear-cut epidemiological data that the virus was not casually transmitted. Nonetheless, the very fact of virus in saliva was treated as sensational news in the media, and our cautious interpretation of the finding was lost in some press speculation about what new risks were now to be expected. I found myself wondering whether we could ever again decide to publish results of an AIDS study (when there were practical implications) without first considering the emotional, social, and political consequences of the publication. But all in all, the scientific developments were exhilarating; the jigsaw puzzle yielded its full view, and we could now understand the epidemic.

All these important steps had been accomplished by the end of 1985. Moreover, viruses related to HIV were discovered in some African primates by several groups, particularly by Ron Desrosiers and other investigators at the New England Primate Center in collaboration with Max Essex, Phyllis Kanki, and their co-workers. Essex also provided serum antibody data showing that another human retrovirus, called HIV-2, was present in West Africa. Essex and Kanki used the newly isolated monkey virus, simian immunodeficiency retrovirus (SIV), in tests of people's sera, and found that some people in West Africa, particularly prostitutes, had antibodies that strongly reacted to SIV. The pattern of reactivity was not like that of people who were infected with the AIDS virus (HIV), so they concluded it was another human retrovirus, one related to HIV but also to SIV. They believed they had isolated a virus from one such person, but later evidence indicated their isolate could have been a laboratory contaminant of an SIV isolate.

Nonetheless, Essex's basic notion and discovery were correct. Indeed, there was another retrovirus, one that appeared to be about 50 to 60 percent different from our HIV—just as HTLV-2 was found to be about 50 percent different from HTLV-1. Isolates of the new virus soon came from Montagnier and his group, and then from a group in Stockholm led by Gunnel Biberfeld and Erling Norrby, and later from Daniel Zagury and my group, and from Reinhard Kurth in Frankfurt. There were now two AIDS viruses, and each one showed significant variations from isolate to isolate for different infected people.

A major difference divided scientific opinion at the time. The Pasteur group obtained their isolates from sick people, and they developed and patented a test for their virus as HIV-2 (or LAV-2). But they looked for their virus mostly or only in people who had immune abnormalities but were negative for HIV-1. In contrast, Essex noted that while it was rare to see HIV-1 in West Africa—it had not yet reached the area—it was common to find HIV-2 there. Moreover, he noted that HIV-2 virus was often found in elderly prostitutes. He reasoned that because the prevalence in prostitutes was so high, HIV-2 was probably, like HIV-1, a sexually transmitted virus. Since it was found predominantly in elderly prostitutes it was not likely to be a new—or, at least, not a very new— infection of West Africans; and since there was little or no AIDS in West Africa, how much of an AIDS virus could this be? But in Europe the press was already announcing many new kinds of AIDS viruses (not only on the basis of this second AIDS virus, but also on the confused premise that different variants of the first virus meant different viruses). The Pasteur group and its collaborators maintained that the second virus was likely to play an important role in AIDS.

The truth seems to lie somewhere in between. HIV-2 has not spread with anything like the efficiency of HIV-1, and it is now clear that it has not produced disease at the same rate as the first. On the other hand, it does appear that it can sometimes cause AIDS, and an analysis of its genes shows that its molecular organization is very similar to that of HIV-1. What fine points of distinction make this virus spread less rapidly than the first, and why is it so much less efficient in causing AIDS? The questions remain unanswered but important. If such differences are found and understood, I think that information will be the real legacy of finding HIV-2.

In August 1984, more support for this kind of virus as the key to AIDS came from Jay Levy and his collaborators in San Francisco. They had a few detections of HIV and were able to get one to infect the HUT 78 T-cell line.

In September 1984 I received a telephone call from Duke University's Dani Bolognesi, an expert on the envelope of retroviruses. Bolognesi had worked with Werner Schaefer at the Max Planck Institute in Tübingen in the early 1970s, and their group had established itself as foremost in the study of the structural proteins, particularly the envelope, of mouse retroviruses. The Max Planck Institute scientists were also known for the practical application of their work. For example, they had obtained evidence that it was the envelope of these viruses that induced neutralizing antibodies in mice, which suggested that this protein would play a critical part in developing a vaccine to any retrovirus.

I had encouraged Bolognesi to get involved in human retroviruses shortly after the discovery of HTLV-1. In fact, he and his Duke immunologist colleague Barton Haynes were the first in the United States to find and isolate HTLV-1 after our reports. But Bolognesi devoted only a portion of his time to work in human retrovirology. With the viral link to AIDS established, I knew we needed his expertise and told him so. In my opinion he was uniquely capable of making a critical contribution to vaccine development, especially in the early stages. Bolognesi attacked the problem with vigor, and in 1984 and 1985 we established an informal international effort to develop a vaccine against the AIDS virus. This group, called HIVAC (for HIV vaccine), is still cooperating in the sharing of ideas, reagents, and information.

On this particular day, however, Bolognesi was not calling me about vaccines. Since we had HIV in permanent cell lines and we now understood its genes, and the steps in their replication cycle, fairly well, we were in a position to think about and to test various chemicals for their ability to inhibit the virus at specific steps.

Duke University is located near Research Triangle in Durham, North Carolina, which consists of a major parklike area housing some large pharmaceutical company research enterprises. One of these is Burroughs-Welcome. Bolognesi knew one of its scientists, David Barry, and both men had become interested in some of the many drugs Burroughs-Welcome had sitting on the shelf. One, in particular, developed as an anticancer agent by Jerry Horowitz at the Michigan Cancer Foundation under an NCI grant, had failed in chemotherapy of cancer many years ago, but Barry saw to it that it was included among the agents worth a try as an inhibitor of a retrovirus. In fact, in the early 1970s, Wolfram Ostertag, working at the Max Planck Institute in Göttingen, had already shown that this agent inhibited the reverse transcriptase of an animal leukemia–causing retrovirus.

Bolognesi asked me to visit and discuss these matters with him,

Barry, and Samuel Broder, then director of NCI's Clinical Oncology Program. The eventual outcome of this meeting was the development of AZT, the first agent proven effective against the AIDS virus in the laboratory, and the first to produce diminished mortality and prolongation of life in symptomatic patients. The work began as a small collaborative study, but it was Broder and his two young associates, Robert Yarchoan and Hiroaki Mitsuya, who effectively brought it through the basic research and into the clinic.

Though not the final answer to AIDS, it was a critical first step in therapy. Perhaps more important, it changed a long-standing presumption in the pharmaceutical industry about what could and could not be done to treat viral disease. Because antibiotics were ineffective against viruses, viral diseases had traditionally been treated for the most part palliatively and by trying to defend against secondary bacterial infections. Now the industry realized that it might be possible to treat viral disease directly, even so serious and mysterious a viral disease as AIDS. This may turn out to be one of the important spinoffs of AIDS research.

It has been said that this period—roughly 1983 to 1985—marked the most rapid progress in medical history in obtaining fundamental knowledge about a new human disease, and that this progress was so rapid because of past basic research, particularly in immunology and molecular biology (especially the latter). Francis Crick, for example, makes this point in *What Mad Pursuit.* I agree with this sentiment, but only up to a point. No one close to the facts can deny that both basic and applied cancer research, directed toward finding and analyzing viruses that cause human disease, played roles at least equally important. The long-gone Virus Cancer Program of John Moloney and the NCI gave us many of our tools and handles in retrovirology. Until it was abolished in the late 1970s it, more than any other program, built up the field of animal retrovirology, later vital in understanding retroviruses in humans.

My own earlier work on the HTLVs and on the T-cell growth factor (IL-2), for instance, did not come solely out of basic research. It also came out of ideas put to the test, ideas designed to find such viruses in humans and to link them to human disease. In turn, experiences with these earlier human retroviruses gave us the necessary background, knowledge, and credibility to propose a retroviral cause of AIDS and an outline of how to approach the problem. IL-2 was essential in growing T4-cells. Contributions from basic and applied immunology allowed clinicians to determine that T4-cells were depleted in AIDS, giving us a critical marker of the disease as well as an opening insight into the mechanism of the disease development.

Applied research in cancer treatment, such as the supportive care programs at NCI, the drug-development program at NCI, and the development and use of multiple drugs (to diminish toxicity and reduce the probability that the patient will develop drug resistance) are all additional areas that technically, logistically, and/or conceptually have already—or will soon—become helpful to AIDS research, the care of people with this disease, and the administration of its programs by the U.S. Health Department and, in particular, by the NIH AIDS administration, now guided so well by Tony Fauci. One could begin both to feel some pride in the advances in research and to focus on the remaining large scientific issues, but this train of thought and work would soon be interrupted by a new complication to the lives of modern scientists.

In late 1983, despite the fact that they had not even positively identified LAV as the cause of AIDS, were unable to culture it in a cell line, and had no blood test result linking LAV to AIDS, Montagnier and co-workers had filed for a U.S. patent for an AIDS blood test. They also filed for a European patent but their request had been rejected out of hand; under the appropriate patent law, such requests must precede publication of the patentable find in a scientific journal. In the United States, a patent can be filed up to one year after journal publication.

We were not aware that the Pasteur group had filed for a U.S. patent. In the thirty years I had been a scientist, I had never filed a patent on any of my scientific findings and I had no thought to do so now. But shortly after we announced our findings and the availability of a blood test, I was approached by officials of the NIH and the Department of Health and Human Services (HHS) and asked to patent our blood test (for reasons I will recount in the next chapter). The way the law stood at that time, I had no expectation of personal profit from the issuing of the patent. I certainly had no idea that the law might be changed one day to allow scientists to earn a limited amount of money from patents taken out in their names. The decision of my administrators to have me apply for a patent for the AIDS blood test would once again drag me out of the laboratory and into the public arena.

11

The Blood Test Patent Suit: Rivalry and Resolution

The prototype blood test was available in our laboratory in early 1984 and approved by the FDA for clinical use in the United States and put into effect one year later. Even starting with all the relevant technology that we had developed, it took that year for the various commercial concerns to make the test practically available to the blood banks. Even so, health officials tell me that this was the fastest pace they knew of for bringing a laboratory procedure to the clinic and to all blood transfusion centers. A few companies were competitively selected for manufacturing the test and were given nonexclusive government licenses to do so. Each had a slightly different approach. All the tests relied on the ELISA technology. Even though it was evident from our results that false positives could, and unfortunately did, occur when transfusion centers used this test, at least one other method for detection of antibodies was used to standardize and confirm the test results. Most based the confirmation test on the Western Blot, as we had done in the lab, to ensure that they would not miss any positives. (See page 183 for a description of these tests.)

Soon it was apparent that the test was preventing a much greater spread of the virus in the world. It also allowed health officials to obtain their first possible knowledge of where, when, and among whom the causative agent was spreading. From that precise moment when the blood test became available, AIDS became a problem that could be measured and scientifically evaluated. Of course, we felt happy about these things, but there were several negative consequences that emerged almost at the same time.

First were those from the scientific-medical side. In early 1984, we did our first tests of Tokyo hemophiliacs and we found that they were uniformly negative for signs of the AIDS virus. A few months later, a substantial percentage had become infected because they continued to use U.S.-supplied blood or blood products, at that time still untested. So, we knew that despite laboratory progress much more was needed from the logistical, practical, and political sides to save lives. Also, I knew that the response to the blood test would not be uniform in different countries. In my experience, the Canadians, Australians, and Swedes were exemplary. Even prior to the publication of our May 1984 papers, their health scientists came to our laboratory, acquired the technology, and quickly brought it home. They were soon ready to protect their blood supply at least as well as we could in the United States. In contrast, scientists in a few other countries did not see the urgency of adopting our test. I remember that a few European medical officials seemed bent on making their own blood test rather than using ours. They succeeded—after a while.

A second difficulty arose from ill-informed pronouncements about the test made by a few local and state health officials. Unfortunately, because of the newsworthiness of any statement on AIDS, these made the national news and then became part of the shared national misinformation on AIDS. The most consistent and destructive were those statements to the effect that no one knew what the test meant—for example, that since the test was based on an immune response to the virus (antibodies), it could indicate prior *exposure* but not infection. Some even claimed it might indicate protection against the disease!

We knew well what the test meant. A significant antibody response to a retrovirus—sufficient to give readily detectable antibody levels in serum—meant a retrovirus was replicating in the individual. Because retroviruses integrate their genetic information upon entry, the immune response to infection shows up too late to serve as protection against it. Experiences with animal retroviruses had long ago taught us this principle. More to the point, our experience with the HTLVs told us the same. In addition, we found in our empirical studies that we could generally detect virus in the blood of people whose sera contained antibodies.

A third problem was the concern about the test raised by organized homosexual groups. They feared, with some justification, the use of this test in an unfair and discriminating way. To hear that no one knew what the test meant greatly enhanced the emotional and less prudent aspects of this concern. And then there was the unfortunate reality that the first real progress science had made in AIDS research for practical benefit seemed designed to protect not the group who was at highest risk but the community at large. Obviously, it could not immediately help those already infected. The argument went that we had created a *cordon sanitaire* protecting the "innocent" heterosexual community from the "gay disease." Though this charge was so ugly it had to be whispered rather than shouted in large-type headlines, it was exploited by those who wanted to strike out at government scientists they saw as having failed to do anything to protect the most stricken group in the general population. Though at first glance it is indeed true that the blood test does not seem to help the infected, the reality is that it was the seminal advance that should ultimately help infected people the most, or at least it was essential for a beginning.

The blood test allowed us to gain fundamental insights into the epidemic, essential for a rational approach to the control of HIV. Of course, the methods for cultivating the virus also gave us opportunities to test drugs for anti-HIV effects. Except for studies of drugs that control opportunistic infections, this was an advance that had to occur to help infected people and it addressed the core of the problem.

Patent is a relatively new word in the biomedical scientific world. As I have indicated, before 1984, my co-workers and I, like most scientists at NIH, had never filed a patent. There was no advantage to doing so, and frankly we would not even have known how to go about it. For instance, we never patented our discoveries of IL-2, HTLV-1, HTLV-2, certain human oncogenes (for diagnostic tests), or the HL-60 cell line. IL-2 is now extensively used by immunologists and clinicians in the therapy of some cancers; a blood test for HTLV-1 is now required prior to blood transfusion; and HL-60 is one of the most frequently employed human cell lines—all profitable to many others without being patented. But we were told that the AIDS virus blood test was a special case, and that we must go forward with a patent. Since then we have patented or attempted to patent most new discoveries that we think might have useful applications.

Lowell Harmison used to be a wide-ranging health department staff member involved in technology transfer. He was known as tough, sometimes rough, very energetic, and capable of solving a problem. I got to know him through the practical application of the blood test. It was his job to take the viruses and cell lines from my laboratory and get them into industrial mass production. Simultaneously, industry was to develop the blood test based on our product and prove its ability to duplicate our results on a massive scale. Essentially, companies were to develop the test for the whole country, in particular for blood banks, and probably for much of the rest of the world as well.

Peter Fischinger, associate director of NCI, Harmison, and the health department officials told me a government patent was needed in order to bring the larger and more qualified companies into the testing. Nonexclusive licenses could then be used to provide incentive for industry to make the sizable investments needed. At the same time licensing would give the government a needed tool in excluding incompetent or even fraudulent groups who might want to start offering the test to the general public.

A government committee would make the selection based on a review of company applications. We were asked to write the essential techniques and to submit virus and cell lines. Our patent was not based on one virus or one cell line but on the process of mass producing the AIDS virus and using it to screen blood for antibodies to the virus. Proposals came to the department from twenty-six companies, from which five would eventually be selected.* Neither my colleagues nor I served on any consulting or administrative committees for funding or for research evaluation of the companies that made the bids. None of us even saw the proposals from the various companies. Our role in the blood test was now over. We were to return to basic laboratory investigations, which we were all too happy to do.

At this stage no one knew that a patent might be of significant financial benefit to a government employee. A few years after the submission of our patent, President Reagan made it legally possible for government workers, for the first time, to receive a small percent-

*Abbott was the first. It was followed by the Dutch company Organon-Teknika and a small U.S. firm called Electro-Nucleonics. Du Pont (with help from a small firm called BioTech) and Baxter-Travenol with Genentech followed.

age of the royalties from their inventions, up to a maximum of $100,000. In the case of the blood test for AIDS, this turned out to be a significant sum.

We later learned that the Pasteur Institute had also filed a patent some months prior to ours, which differed from ours in that its stated aim was a test for AIDS rather than a blood test designed to protect the blood supply. The French test had three problems: (1) only 20 percent of AIDS patients' sera scored positive in the studies they had reported and in the studies they submitted in their patent; (2) they could not find antibodies to the envelope of the AIDS virus, only to the core, and detecting antibodies to *both*, but especially to that component of the envelope called GP41, is a key to a successful blood test; and (3) they had not produced the virus in a cell line; consequently, the amount of virus was very limited, and any commercial production would have to devise ways to pass the virus every few days into a fresh sample of a newborn's umbilical-cord blood. No large-scale blood test was really possible. Also, they had not supplied compelling evidence for the virus as the cause of AIDS, nor did they attempt to argue the case, though I do not know whether this is important in patent law.

The arcane world of government patent offices is not one for any of us, scientists or otherwise, to understand. Our patent was issued. The Pasteur Institute's at first was not, which enormously intensified the problems between our groups. In an effort to reach a reasonable solution, I met with François Jacob, Nobel laureate and president of the Pasteur Institute, over a long dinner arranged by my friend Daniel Zagury of the University of Paris sometime in late 1985. It was held at the home of the then science editor of the newspaper *Le Monde,* Claudine Escoffier-Lambiote, who also served on the Board of the Pasteur Institute. In the United States, such a dual allegiance would hardly be possible. It would be as if the editor of the medical section of the *New York Times* or the *Washington Post* were to sit on the main board of NIH. The dinner was light and delicious, the conversation less so.

Jacob is one of the great figures of twentieth-century biomedical research and a man I hold in the highest esteem. He believed that it was reasonable for the NIH and the Pasteur Institute to share the royalties. I agreed. The Pasteur Institute did not have much money, he explained, and at least a third of the royalty money would be appropriate for the role of their scientists in describing the correct virus first (by then it was clear that they were all of the same virus type). I agreed. He also thought that as the central person, I should and could intervene to see that this was done. I disagreed. I did not think it would be possible for me, a U.S. government scientist without administrative authority, to take on such a role. He told me that unless this kind of fair sharing of the royalty money was worked out, they would have no option but to pursue legal action and, since I was the only identifiable person in the U.S. government related to this patent, it would not be a happy position for me. He could do nothing once things were in the hands of litigation lawyers and others. I discussed this with the then NIH director, James Wyngaarden, and NCI director, Vince DeVita, and also with health department officials, who argued, as I had expected, that the matter was not in my hands. I do not complain about their opinion. They were right.

The lawsuit began, and unfortunately Jacob was also right. Public relations firms were activated and distortions began appearing. On and off for the next two years, my life would be overtaken by lawyers and tension. I read scandalously incorrect accounts in European newspapers of how NCI scientists were making money directly from Abbott, Du Pont, and the other companies carrying out the blood test. These articles were written before the change in the law allowing a government worker to earn such significant royalties and, in any case, did not reflect the correct amounts or the fact that the fees did not come from companies.

As it became apparent to me that these legal squabbles could go on endlessly, I began to seek advice from several quarters. At the same time our scientific work was being intruded upon by what should have been a purely legal matter; but in the climate that existed, we could hardly have the luxury of focusing solely on science. Even relatively inconsequential errors became grossly exaggerated examples of our perfidy. For example, we had prepared a total of three papers for *Science.* They included an electron micrographic picture of one or two of our isolates. But Jörg Schüpbach had carried out a considerable amount of the protein work, and he was critical to our development of the Western Blot for clinical use in the HIV blood test. He reasonably argued for a fourth paper, to be published at the same time as the other three, describing some of his protein work, on which he would be the first author. In deciding to submit this fourth paper, we also, largely for illustrative purposes, added a composite photograph that showed the new HTLV-3 retrovirus alongside photos of HTLV-1 and HTLV-2, in discrete stages of their life cycles. To show the series required selecting frames for the composite figure. Inadvertently, one of the frames used for the composite was a photo of the LAV sample we had temporarily grown.

I did not have any way of knowing this, since we received only the final composite—which was made from a photograph of mounted pictures of the virus at various stages of its budding and formation—and since different isolates of the virus are structurally the same and visually indistinguishable. But the microscopist later discovered that the composite had been made in part from a picture of LAV.* When it was brought to my attention, we, the authors of the paper, wrote a letter to *Science* explaining the problem. In the prevailing atmosphere, that proved to be a costly mistake. Earlier I had openly expressed my concern about the behavior of the British magazine *New Scientist* in going to press with our results prior to their being published in *Science,* which resulted in the rushed press conference. *New Scientist* now took the lead in writing a nasty article. John Maddox, the editor of *Nature,* and some British virologists, including Robin Weiss and William Jarrett, intervened with strong supportive letters and with harsh words about *New Scientist*'s policies, but the journal refused publication of the most important of these.

The French group then reported on an apparent scientific advance, but in reality it turned into a temporary setback. They reported in July 1984 that they too had succeeded in getting LAV to grow in culture, but the cell they used was a B-cell line. This was unexpected news, because neither HIV nor any human retrovirus to date infects B-cells very well; it would indeed be a strange source for continuous mass production of the virus. They also reclassified their virus from a type-C to a type-D retrovirus. We noted that the human B-cell line they selected (known as BJAB) had been co-cultured with an EBV-transformed cell line. But we also knew from past experience with this particular EBV cell line that certain stocks from Europe were contaminated with type-D squirrel monkey retrovirus. Montagnier was very

*This had been done by mistake. When Mika sent his work on cell culture with LAV for electron microscopic evaluation, he properly labeled it LAV. Apparently the technician at the contract company that performed the evaluation selected LAV for one of the pictures, not knowing it was the virus from the Pasteur group. It had simply been filed under a general title of AIDS retroviruses sent from my lab. At that time the name LAV had not yet appeared in the scientific literature.

likely producing this virus, perhaps with much smaller amounts of his LAV strain of the AIDS retrovirus. Later Robin Weiss succeeded in getting LAV to grow in a human leukemic T-cell line called CEM, one of those we had already succeeded in infecting with HIV. In 1985, he made it available to Montagnier, who subclassified the Pasteur group's isolates as lentiretroviruses, a point we now all agreed on.*

In the summer of 1985 I met with Jean-Claude Chermann in Daniel Zagury's home in Paris. Zagury is an immunologist at the University of Paris, and in particular he is a student of the T-cell. We had been close friends for several years, ever since my former associate Doris Morgan moved to his laboratory and began studies with Interleukin-2 there. Zagury was also a friend of Chermann and badly wanted to see the problem settled. As noted earlier, I had also known Chermann well for about fifteen years. Differing greatly from Montagnier in temperament and style, Chermann is more emotional and open. Like Montagnier, his expertise, and that of his longtime close associate Françoise Barré-Sinoussi, is in cell culture and the interactions of virus and cells in culture. But he also had expertise in retrovirology. Years before, he had trained in Bethesda at the National Cancer Institute working on mouse retroviruses with Peter Fischinger, who had since been named associate director and worked with DeVita.

Within about forty-five minutes, Chermann and I wrote a history of the scientific events leading to the discovery of the AIDS agent. We had almost no differences of opinion. If I had dealt solely with Chermann, I think we could have solved the whole thing in one day, but there were other forces on both sides. This written history was never used or published because of a request I received from a member of the Pasteur Institute Board, the same Claudine Escoffier-Lambiote. Since there were other efforts being made to resolve the dispute at the same time, I complied with her request, but I have regretted doing so. Had we made that history public, it probably would have led to an agreement one year sooner. The other efforts proceeded from several directions. They included discussion among embassy scientific attachés and then with me; the entry of William Walsh, director of Project Hope, into behind-the-scenes meetings; frequent advice and

*Weiss would later relate to me that Montagnier told him soon after that the CEM cells Weiss had given him were contaminated with mycoplasma. Therefore, he (Montagnier) had to reinfect the cell line himself. This also meant that Weiss and his institute would not share any patent royalties should they be forthcoming to the Pasteur Institute.

help from Nobel laureate Howard Temin; and above all, the almost missionary zeal and Kissingeresque shuttle diplomacy of Jonas Salk.

All of us knew Salk from polio vaccine stories and from brief meetings. I was astonished by his tenacity and focus on our problem. Perhaps these characteristics were also his chief strengths in his work on the polio vaccine. From meetings in plush hotel rooms, such as Washington's Ritz Carlton, where Salk made his Washington home, or in Paris cafés, to discussions in the more barrack-like side rooms at scientific meetings almost anywhere in the world, Jonas Salk dedicated the years 1985 and 1986 to bringing harmony among the institutes, laboratories, and especially individuals involved in this conflict. With an obsessive commitment, he catalyzed discussions between Montagnier and myself and pushed us to write, rewrite, and refine a series of statements on the scientific history of the AIDS virus, the completion of which Salk believed would ignite an agreement. But that was only a fraction of the problem. Behind it lay the legal and institutional agreements, which somehow moved forward. As the scientific history progressed, it was paralleled by embassy diplomacy and the dedicated work of two U.S. health department lawyers, the chief counsel Ron Robertson and in particular his associate, Robert Charrow whom, I suspect, along with their counterpart, Ira Milstein, representing the Pasteur Institute, masterminded the agreement.

The annual international conference on AIDS in June 1986 was held in Paris. Productive discussions had been progressing reasonably well. My own feeling was that the issues of disagreement were small, the arguments stale and of little relevance to the scientists on either side. The scientific problems were the issues to keep our minds on. Nonetheless, because of massive media interest in the science of AIDS and in the U.S.-French problem, tension was high. Following my address I was deluged by the press with questions about both aspects. Salk played one more of his diplomatic roles by keeping me hidden in the men's lavatory until my adrenalin level returned to normal and the press went away, and then arranging a planned discussion with reporters in a manner much less likely to result in an emotional response from me.

Four months later my colleagues and I were back in Paris, this time for the annual meeting of the Pasteur Vaccine Company. Throughout the two-year patent dispute, scientific interchange never slowed, so it was not unorthodox that we were attending this meeting. Scientifically sound information was presented but, as in previous and, until very recently, also subsequent meetings on vaccines, no major progress had been made. I was disappointed. But I was not disappointed at the changing atmosphere. It wasn't just the beauty and warmth of an autumn day in Paris that signaled the difference; the mood was decidedly one of exceptional friendliness. Added to this, in their annual awards the Lasker Foundation and some others would shortly give equal recognition to the French and the U.S. teams.

The next time I met with Montagnier was March 23, 1987—my fiftieth birthday. I was on my way home from a brief meeting in Europe, and I stopped to visit my old friend Prakash Chandra, head of molecular medicine at the University Medical School in Frankfurt. I was there to discuss some antiviral chemicals with him, as well as to celebrate my birthday with friends from Germany. This provided a bonus: the opportunity to spend a few hours with Montagnier writing the final aspects of the scientific chronology of events, which was subsequently published in the British journal *Nature.* We also came to an agreement on the generic name for the virus. Human immunodeficiency virus, HIV, had already been recommended by others as an appropriate name that recognized the primary role of this retrovirus in the cause of AIDS. Soon the French premier, Jacques Chirac, visited the United States, and he and President Reagan announced the agreement.

The immediate effect of the agreement was that some percentage of the royalties received from the blood test fees was allocated to the Pasteur Institute and to NIH for research purposes. The remaining money was channeled to the new World AIDS Foundation where Wyngaarden, the Pasteur Institute's Director Raymond Dedonder, Montagnier, myself, and the lawyers (Robertson and Milstein) maintain Board positions. The objective of this foundation is to direct these funds toward AIDS-related work in needy countries. To a substantial degree, initial projects will probably involve help in testing blood for HIV.

In the summer of 1987 the American Federation for AIDS Research (AMFAR) held a fund-raising event and succeeded in garnering the participation of President and Mrs. Reagan and Elizabeth Taylor. In France a gala fund-raising event for AIDS took place to coincide with the signing of the agreement. Mrs. Chirac, Princess Caroline of Monaco, the English actor Michael York, and the singer

Shirley Bassey were among the participants. The French and U.S. scientific teams shared in these events.

In 1988 Montagnier and I wrote for *Scientific American* a more detailed history of the scientific steps leading to the discovery of HIV, its linkage to the cause of AIDS, and the development of the blood test, an article that was catalyzed by *Scientific American*'s editor, Jonathan Piel, and by the work of the energetic and exceptionally competent science writer John Benditt. This was their October 1988 issue, and it was devoted entirely to the science of AIDS. In June the AIDS International Conference had taken place in Stockholm. Bill Blattner notes in his October 1988 *Scientific American* article—in which he provides an insightful, critical, and informative review of a popular book related to AIDS, Randy Shilts's *And the Band Played On*—how the atmosphere in Stockholm was in great contrast to that of the prior two international AIDS meetings. Although even larger and noisier than the ones in Paris in 1986 and in Washington, D.C., in 1987, the Stockholm meeting exuded the feeling of a team working together to solve a problem, the promise of good collegiality, friendship, and cooperation. It was the best moment I can remember in AIDS research.

The old adage that those who forget history are condemned to repeat it is helpful, but dwelling on the past can eradicate the future. So I do not like to think too much about the alternative pathways of the past. Sometimes, though, I cannot help but think back and note where I might have been wrong. Although it can be difficult for one who works for the government—even a scientist at NIH—to know his or her own level of authority, I have little doubt now that I could and should have acted more assertively to control events that greatly affected me.

Yet, I know that things could have been better across the ocean as well. In the end, I am very glad it was productively resolved. Sandra Panem, in her 1988 book *The AIDS Bureaucracy,* gave an appropriate perspective when she said: "Regardless of the *political* [italics in original] settlement concerning who discovered the AIDS virus [and, of course, the development of the blood test, by inference], and who will garner Nobel prizes or public opprobrium, the May 1984 acceptance of HTLV-III/LAV as the cause of AIDS irreversibly changed the nature of managing the epidemic. Prior to that time AIDS research was groping: now it had direction. Scientists could go on to real targeting of specific tests and treatment and prevention strategies. Whereas coordination and augmented resources had always been desirable, they now became mandatory. And so the debate over research management strategies as well as questions of public policy was dramatically changed."

IV

A Scientist's Look at the Science and Politics of AIDS

12

The Alarm

The years after AIDS was first recognized (1981–82) and the years of intense discovery (1983–85) were also years of confusion, misguided hostilities, occasional panic, and the use and misuse of AIDS to promote special interests or prejudices. Despite the advances in our understanding of the disease, and much public education, this kind of activity continues, though to a lesser extent, today.

Once it was realized that AIDS was likely to become a serious epidemic, many people overreacted and tried to place blame. For example, a major political leader in Bavaria, who was being advised by a country doctor, flirted with the idea of developing camps for HIV-infected people. This doctor had lived in Sweden during the post–World War II years and was a self-proclaimed AIDS expert who wrote pamphlets and a book that highlighted an alarmist viewpoint. To spread his alarm he arranged elaborate scenarios, such as noting that although it is claimed that the virus cannot be casually transmitted, what if certain sequences of events occurred in a barber shop? With the help of the best science artists, he dressed up his book with numerous and colorful scientific artistic pictures. He also included some real scientific pictures of cells and viruses, including some he obtained from me. It is no wonder that on a trip to West Germany during the mid-1980s, I heard many expressions of intense fear about casual transmission.

Clear heads and solid science eventually prevailed, as people in Bavaria (such as Friedrich "Fritz" Deinhardt, director of the Max von Pattenkofer Institute in Munich, and Reinhard Kurth, president of the Paul Erhlich Institute in Frankfurt) and scientists in Berlin (such as Professors Karl O. Habermehl, Hans Gelderblom, and Meinrad Koch) fought successfully against this kind of mischief.

I returned to Germany in 1988 to give the Boehringer-Ingelheim lecture series. In East Berlin I attempted, unsuccessfully, to meet a Professor Siegel, who had invented and actively promoted the notion that HIV was created by a U.S. government scientist, with the help of the Frederick Cancer Research Facility in Maryland (it was, in fact, formerly in part a military research area that included studies of germ warfare, but the U.S. government had long ago abolished it). I am told that Siegel's thesis was that HIV is a recombination of HTLV-1 with a lentiretrovirus of sheep known as the visna virus, constructed by recombinant DNA technology. This is an interesting notion that superficially makes tantalizing science sense, since the HTLV component would conceivably provide the human and T-cell targeting capacity, while the visna component would give it its macrophage tropism and its ability to mutate so easily. The problem is that HIV has been proven to have existed long before recombinant DNA technology was invented; moreover, there was little relationship between the published gene sequences of HTLV-1 and HIV, and only a small amount of genetic similarity between the gene sequences of visna and HIV.

The notion was simply stupid, malicious, or both. Still, some people seem capable of believing any nonsense, so long as it reinforces their prejudices or supports their politics. Some of the serious East European press gave it credence for a while, and there were many Americans who believed such wild charges despite the many attempts of scientists there to expose them as absurd. For a long time these stories circulated widely—even among people who should have known better—with dangerous results. Occasionally, my life and the lives of other U.S. scientists or science administrators were threatened. Once while I was lecturing at Brown University a young man holding a box announced he was going to kill me because he had read such things. Similar events occurred at NIH. Worse, such a theory was thought possible or convenient by some African officials who felt themselves to be on the defensive because scientists were beginning to recognize that the origin of HIV might have been in rural central Africa.

For every one of these essentially political diatribes, however, there were scientists from all over the world who cooperated with one another to seek out the truth. One of my favorite episodes of this nature took place in Erice, a very small town on the crest of a mesa on the northwest coast of Sicily. It is an ancient town, with a past of Norman, Roman, Greek, and pre-Greek people. Overlooking the dramatic vista of distant mountains and the flat seaboard of Marsala, at the edge of the cliffs, rest the ruins of a Greco-Roman temple to Venus. This tranquil spot is the setting for a postgraduate course in current topics in physics, whose primary purpose is to foster international interchange. The program also includes some biology courses, and while I was participating in one on lymphocytes, I was asked to give a special lecture to a group that included about ten Russian physicists who were especially eager to hear about retroviruses.

When I finished, they asked me what I thought of the claim that the AIDS virus had been man-made in the United States. My response was quick but conclusive, and their response to mine was joy and applause. One of them (I believe it was Professor Eugene Valikhov of Moscow) promised me he would bring the problem to the attention of key people who knew the new Soviet leader, Mikhail Gorbachev. I do not know whether this man was responsible for what happened a few weeks later when an official Soviet pronouncement condemned the notion of a man-made virus in the United States or anywhere else.

A second positive outcome occurring at this meeting came out of a televised press conference. I was asked almost the same question by a reporter, but in a different way: "*Could* the AIDS virus be used for germ warfare?" I responded that essentially this was impossible, that no country could control it. I added that the AIDS virus would instead help force openness and collaboration in science among nations. It would help bring people together. Antonio Zicchichi saw the interview on television while in his Geneva home. Zicchichi is a physicist, and he is responsible for the Erice conferences. He is also the director of a major international program called the World Lab, whose main objective is to bring people together on beneficial scientific projects, particularly projects that bridge east and west or north and south.

Immediately he decided that a virus project, particularly one designed to help in Africa, should be funded and asked me to help with its direction. He would provide substantial funds and my role would be advisory and not too time-consuming. I accepted. Within a year we had initiated twelve projects run by European scientists with contacts to areas of Africa in special need of study and help.

Panic, sensationalism, and overreaction to AIDS seem to have become part of our response to the fear generated by this disease. For example, there is the potential for serious harm in the recurring discussions of casual or insect-borne transmission of HIV. I have already mentioned the reaction that resulted from our finding low levels of virus in saliva. A second and more persistent one has to do with mosquito transmission of HIV, especially in Belle Glade, Florida, where schools refused entry to HIV-positive children. Belle Glade indeed had an HIV-positive cluster. Belle Glade has mosquitoes. Investigators from the Centers for Disease Control demonstrated that the Belle Glade phenomenon could readily be accounted for by known routes of HIV transmission such as drug abuse, which was prevalent.

But a few reporters and one or two scientists were inclined to stretch their incomplete knowledge to support apparently logical conclusions. HIV was transmitted by blood—this was the fact. Therefore mosquitoes, which drink blood, are transmitters—this was the erroneous conclusion. Phil Markham, my colleague, in collaboration with some insect biologists, who were predisposed to believe this theory, found that unlike the parasite of malaria, HIV does not replicate in mosquitoes. Moreover, mosquitoes that fed on infected cells in laboratory cell culture systems were not observed to transfer the virus to uninfected cells with which they were brought into contact. These data supported the careful epidemiological studies of CDC workers that had already indicated that HIV was not insect-transmitted.

Nonetheless, one evening my younger son Marcus came to tell me that I was on national television on the evening news, warning America about the dangers of mosquitoes. I saw on the screen a silent tape of a lecture I had given months earlier on anything but insects at the International AIDS Conference in Washington. The newsman was doing the talking, stating that I had made an announcement, and there I was with my mouth moving—a viewer couldn't expect any clearer proof than that. I am told that the story emanated from a cocktail party where a collaborator inclined at the time to believe in the probability of mosquito transmission told a local reporter of my being involved with him in the mosquito work. He neglected to tell the reporter of the negative results we were obtaining. From reporter to reporter to national TV within the day, all without any contact with me! Although this was one of the more sensationalistic and incompetent TV performances, it was not unique. Between 1984 and 1986 there were others like it, and I know of no explanation except the pressure to fill daily reports in a burst of seconds and not to feel left out if another reporter has said something that might shake a few viewers out of the hypnotic trance in which most people watch television news.

Unfortunately, after the mosquito scares had subsided, they were thrust forward in the headlines once again. This time it was because of some unusual results reported by Jean-Claude Chermann of the Pasteur Institute. At the 1986 Pasteur Vaccine Company meeting in Paris, he had described finding DNA sequences of HIV in many African insects—not just mosquitoes but also bedbugs and cockroaches! The procedure was unusual. For several years, scientists had used a certain procedure called a Southern Blot when used for DNA (named after Ed Southern of Edinburgh* who first developed the assay) and a Northern Blot when used for RNA (named so for obvious reasons). Analogous to the later-developed Western Blot, which separates proteins and tests for antibodies to a mixture of these proteins, Southern or Northern blots subject *purified* DNA or RNA of a sample present in different sizes to a process that separates them. These nucleic acid fragments are then transferred to a special paper by blotting. Then a probe of the test material is used in a procedure developed earlier by other scientists and called molecular hybridization.

The probe can be made radioactive. If the probe is HIV and if there are sequences of HIV in the sample, the HIV probe will stick to its complementary sequences (*positive hybridization*). These are visualized by using a technique called *radioautography* in which an X-ray film is applied over the paper. Wherever the radio-labeled probe has hybridized (and not been washed away), it will leave its signal on the film. In other words, the method can tell whether there are HIV sequences in the sample and the size of the fragments it appears in.

Chermann used a modified version of this method, in which a whole

*Now professor at the Waltham College, Oxford University.

insect is "squashed" on the paper and the probe incubated with the entire thing. With a bit of humor, he called it "squash blot." For an insect to have viral DNA required that HIV infect an insect and actively replicate in the organism (for the RNA to be converted into DNA). If cockroaches do not bite humans, how did they get infected? Chermann believed it was by ingestion of human excrement. This would mean that these insects not only take in blood (or cells with virus in some form) but that they then get infected, that is, that the virus must manage entry into insect cells and conversion of viral RNA to DNA—at least part of the HIV cycle of replication. But this would be odd because, at low dose, HIV targets few cells even of humans.

Even if these results are not artifacts, they show only that the insect may take in virus, not that it transmits virus to something or someone else. But I think the explanation lies elsewhere. I suspect that these insects contain *cellular* genetic sequences related, but not identical, to HIV. Could they be the sought-for ancestral cellular origin of HIV—as endogenous proviruses?

Heterosexual transmission of the AIDS virus was obvious from the beginning—obvious because such transmission occurs with HTLV-1 and HTLV-2, significantly less infectious retroviruses than HIV; obvious from the epidemiological studies in Africa; and inherently obvious because no virus could have remained within the gay community throughout history. After the initial difficulty in getting this point understood, however, the *degree* of heterosexual transmission came to be vastly overstated in certain quarters, most egregiously in a widely publicized study by Masters and Johnson which was limited in its size and scope. Such severe fluctuations wear down the credulousness of the general public.

Of all the reactions and counterreactions that have occurred in the brief history of the AIDS epidemic in the United States, none competes with the issue of blood testing. I have already discussed some of its early history and problems. Later there were repeated debates about the who, when, where, and how of tests for HIV. We do need to follow the epidemic, and for this we will need regular and extensive studies of our population. But as medical scientists have repeatedly warned, forced testing will drive away those we most need to test. We have also noted the unfairness of discriminating against people infected with a virus that in the Americas and Europe was most often transmitted by intimate homosexual contact, the mere suggestion of which can destroy lives, careers, friendships.

Further, scientists have repeatedly pointed out the stupidity of barring the entry of AIDS patients into the country. The HIV-infected, but currently healthy, person—who makes up the vast majority of those who carry the virus—will not be barred, and this person will be more likely than one already sick with AIDS to be still sexually active and a potential agent for new infections. Most important, persisting in this kind of policy will earn for us the international disrespect it deserves.

It is important to consider the epidemiological facts and proceed from them to a discussion of the more basic issues of the biology of the AIDS virus and how it produces its multiplicity of effects. I would like to do this by raising and answering here the thirteen questions I am most frequently asked.

1. *What is the extent of the disease in the world?* This and most other questions of numbers cannot be answered with absolute precision. Underreporting and problems in diagnosis make the figures epidemiologists work with less reliable than they would like. Problems with diagnosis easily occur when, for example, a person is in a tuberculosis-endemic region and dies of tuberculosis activated by HIV, in which case the cause of death may be listed only as TB. The proximate cause of AIDS deaths, and the cause most easily recognized by local pathologists, is one of the local microbial flora.

Our best guess is that there are close to 100,000 cases of AIDS in the United States and probably about 1 million worldwide—though the *reported* estimate is about 300,000 worldwide.

2. *What is the extent of the infection now?* Again the numbers are imprecise, far more so than the number of AIDS cases. The estimate of 1 million infected people in the United States seems reasonable, and I would think the total worldwide must be about 10 million.

3. *What can we expect its extent to be in the future?* Epidemiologists and other health officials are almost forced to take a crack at this question, often with mathematical analyses. I don't trust all mathematical approaches to biological issues. They are often wrong because it is

difficult to know all the variables. We should be able to make some reasonable estimates of the future number of AIDS cases if we know the number of people infected, the time between infection and symptoms of AIDS, and the constancy of virus and co-factors. But as we have just seen, we do *not* know the number of people who are infected worldwide, although we do have reasonable estimates for the United States, Europe, and parts of Asia.

As for the time from infection to development of AIDS, the estimate has changed many times. At first it was a couple of years, then four to five, then seven, and now the estimate is that in about ten years better than half of the infected people will get AIDS and the majority, symptoms of some kind. Only a few years ago a careful mathematical analysis published in a prestigious U.S. scientific journal stated that if AIDS has not occurred within fourteen years after the time of infection, it will never occur. I doubt whether any researchers working on the biology of this virus still believe that, if they ever did. Another important factor is variation in HIV. Perhaps some strains are less pathogenic than others. Finally, the presence of co-factors may vary over a period of time, and this too could change the incidence. Putting all of these concerns aside and dealing with the most abundant data, the CDC estimates that in 1991 alone there will be 74,000 *new* AIDS cases, at a cost of $8 billion dollars to care for and treat these patients. In that year, about 54,000 deaths from AIDS are foreseen.

More important for the future epidemic is not the number of AIDS cases but the rate of infection, and this is far more difficult to predict. It depends not only on co-factors or virus variants but also on people's behavior patterns and how these have been affected by educational programs and other factors. I do not believe it can be closely approximated.

4. *What is the origin of AIDS and the AIDS virus?* Since diseases often mysteriously appear, then disappear, it is easier to document the disappearance of a known disease than the emergence of a new one. Part of the difficulty is that the disease may be old but only newly recognized because of poor historical records or inadequate past medical procedures. We must be careful to distinguish between sporadic rare cases and epidemic disease. Like syphilis, which appeared as an epidemic for the first time in 1495 but was probably sporadic earlier, the AIDS virus surely became prevalent for the first time only in the

1970s, at least in the United States and other developed nations, but did exist in humans prior to that first recognition of it.

Many sera obtained years ago from many parts of the world and then stored frozen have now been tested for HIV, and to my knowledge all such tests indicate that HIV has only recently become widespread. On the other hand, there is evidence of sporadic HIV infection within isolated populations in the 1950s and 1960s, and it is probable that it occurred much earlier. Belgian clinicians have described a disease indistinguishable from AIDS in the early 1960s in a few people from rural central Africa seen in the clinics of Kinshasa, Zaire. The earliest specimen of which I am aware is a report of a Norwegian sailor who developed an AIDS-like disease in the late 1950s after a visit to a central east African coastal city. Specimens from him at that period were available and tested positive for HIV. Because of these early case histories; because parts of central Africa now have the highest infection rate; because its closest relative, HIV-2, is all but limited to equatorial West Africa; and because several species of monkey are infected with the related simian immunodeficiency virus (SIV), we obviously suspect an equatorial African origin of HIV-1.* As we have already noted, at least one form of this monkey virus is more than 80 percent homologous to HIV-2, whereas no one has found a monkey virus that has been proven to be more than 40 to 50 percent homologous to HIV-1.

This has led some of my colleagues in the field to suggest that the natural origin and evolution of these viruses is: SIV → HIV-2 → HIV-1. But this is not easy to accept. If it is true, then why don't we find HIV-1 in West Africa? Or at least why did HIV-1 appear rarely and late in West Africa, where HIV-2 dominates? And why is HIV-1 prevalent in central Africa, where HIV-2 is rare? Clearly there is a missing link—perhaps an undiscovered monkey virus closer to HIV-1, which will indicate that there were at least two independent infections of humans by these monkey retroviruses: some SIV to HIV-2; others SIV to HIV-1.† This would not be too surprising since

*Amazingly, in the early part of my research on AIDS (early 1983) I was visited by Ann Guidicci Fettner, a freelance writer who told me emphatically that the origins and epicenter of the epidemic were in a river basin near Lake Victoria. She also stated that she believed the virus came from African green monkeys, apparently due to her experiences and observations in central Africa.

†There have, in fact, been very recent reports of an isolate from a chimp with an SIV closer to HIV-1 than any prior SIV isolates.

HTLV-1 may also have come into humans from African monkeys, evolving from STLV-1. We may never know for certain the answers to these questions, but they are of more than academic interest because answering them may help us avoid future zoonotic catastrophes—that is, the transmission of disease from lower animals to humans.

What is clear is that a once rare and remote virus became relatively common and widespread, almost surely due to certain changes in demography and lifestyle in the twentieth century. We can imagine a stable relationship between host (monkey, chimp, and human) and parasite (HIV-1 and HIV-2). People in parts of equatorial Africa may have been sporadically infected. The first infections of man may have occurred from exposure to the flesh of infected simians (monkeys and chimps). We cannot date this, but some molecular analyses of HIV-1 suggest it may have appeared within the last few centuries. It could be argued that HIV-2 infection is much older, because HIV-2 appears far less harmful to the endemic infected West Africans than HIV-1 is to people in general. The alternative is that HIV-2 was simply less pathogenic to humans from the start and its entry into West Africans also relatively recent. We simply do not know.

We do know, however, that originally the infection must have been contained, probably rural, the virus dying off with the occasional sick person. The apple cart could have been upset by demographic changes. European powers influenced changes in agricultural ways. Sometimes, even though European technocrats may have departed out of the best of intentions (leaving Africa to Africans), some populations that seem long before to have lost their old ways seem not to have been able to keep the new ones learned from the Europeans, and this may have contributed to migration, usually to cities. Consequently, there were new sexual contacts and drug abuse. Surely the rapid rise in international travel due to the airplane must also have contributed to the sudden change in the relationship between HIV and humans. The use of blood and blood products for medicinal purposes since the middle of the century, but until recently without the advantages of sterile equipment, must also have played a part, as did intravenous drug addiction.

5. *Who gets AIDS?* Whoever gets infected with HIV is highly variable from region to region and group to group. The lessons with HTLV-1 are instructive in that we know from the HTLV studies that

we cannot generalize on a region, a village, or an area. What is true in one locale is not true in the next. In the United States and Europe today, as from the start, gay men make up the largest group of HIV-infected people, but intravenous drug addicts are an extremely high percentage in some regions. In the United States about 40 percent of children born of or breast-fed by infected mothers become infected. Hemophiliacs also make up a small percentage of infected people, although now that the blood supply has been protected, blood-transfusion cases of AIDS will gradually decline and should approach zero in most industrialized regions of the world. Heterosexual transmission is more efficient from men to women, especially in the presence of venereal infections that increase the number of white blood cells (leukocytes) responding to the infection, and thus the probability of transmission of HIV. Also, some groups have shown that ulcerative lesions, such as those characteristic of some well-known venereal diseases, especially chlamydia, chancroid, certain herpes lesions, and syphilis, also increase virus transmission and may markedly promote transmission from women to men. The higher prevalence of these diseases in parts of Africa almost certainly accounts for the now far more pronounced heterosexual transmission in that continent.*

Asia, South America, and Haiti also illustrate the extreme regionalization of human retrovirus transmission. For instance, HIV-1 is already spreading in Thailand and destined to be a long and serious problem for that country, but it is not a significant problem in China. Rio de Janeiro may have had one of the highest incidences of HIV in its blood supply early in the epidemic, as indicated by a small survey we carried out early in 1984, but many other regions of South America were essentially HIV-free. We also tested blood from a number of Haitians shortly after we developed the blood test, and the HIV-positives were among the highest we had seen in blood from the early 1980s. But there are other Caribbean islands (for example, Trinidad) where the rate was then much lower. I would think this is best explained by the better medical care and equipment available in Trinidad compared to Haiti, of the kind that ensures less likely contamination through blood (for example, availability of sterile needles and syringes).

Other factors may also contribute to the discrepancy, such as ritual-

*And we never know when and if these venereal diseases will increase in other areas, such as the Americas and Europe. If they do, we can expect easier heterosexual transmission in these regions.

istic voodoo practices. Haiti also illustrates how fantastic and new things occasionally happen to biomedical researchers who work in this field. My co-workers and I were asked (almost begged) by some U.S. medical colleagues working in Haiti who had sent the blood to us not to publish the data; the political climate at that time was such that they believed their lives were at risk.

The essence of the answer to this question, then, is that anyone whose bloodstream is frequently exposed to blood from random persons and anyone in intimate contact with bodily fluids of infected persons, as by sexual intercourse, is, of course, at an increased risk of getting AIDS.

6. *What are the greatest problems with AIDS today?* Most of the attention given to AIDS in the United States has been focused on gay men. This was particularly true in the early stages of recognition of the epidemic. But on a global scale, the greater problems now are the spread of HIV in central Africa, parts of Asia, and probably parts of South America, as well as the global problems of intravenous drug addicts and others exposed to contaminated blood, as in those areas where syringes and needles used for medical purposes are still frequently contaminated—as happens in parts of Eastern Europe.

Babies born to mothers of IV drug addicts or mothers from the areas of high exposure for other reasons are also the problem of the future. At a time of economic difficulties and enormous problems with other microbial and parasitic disease there, HIV-1 is hitting parts of equatorial Africa hardest. Today there are some regions where over 20 percent of randomly selected people on the street, 40 percent of people in hospitals, and almost as many in the military are HIV-1–positive. From the beginning the evidence indicated that the virus was being spread by sexual relations between men and women. The contrast between Africa and other regions in the rate of spread by this means has led to many speculations and some scientific studies.

Frank Plummer and his co-workers in Manitoba, Canada, and their collaborators in Kenya have provided the most logical explanation and the best data. As I noted in the previous answer, their results and those of others support the role of other venereal diseases in HIV-1 transmission, particularly those diseases in which ulcerative genital lesions occur. These lesions are potentially rich sources of HIV in infected individuals. Also, Zagury and I noted that HIV in the semen of infected men was associated with the presence of white blood cells (macrophage and T4-cells), cells that would be much more abundant in a person with a venereal disease, such as gonorrhea. These sexually transmitted diseases are an extremely significant problem in central Africa. To help fight HIV and AIDS in these places requires doing something about these problems, too—unless an HIV vaccine is developed.*

7. *Aren't there today much more serious medical problems than AIDS, such as cancer and heart disease in the developed nations and malaria, TB, parasitic gastrointestinal, and respiratory disorders in Africa?* This is another way of saying: Isn't too much attention being given to AIDS today and too much money being spent on it? And won't this endanger other health and medical research programs? The answers to these concerns are complex—at least to me. The first reaction must be that neither interest nor money invested in AIDS research need be in conflict with interests and funding in other areas of health science. AIDS is a *new, additional* problem. Society should tighten its belt elsewhere—not at the expense of other health areas.

Also, AIDS is increasing and its cause is known. There are things we can rationally do to reduce and someday stop this disease. With greater understanding, there may be steps we can take.With some of the other diseases this is not the case. With tuberculosis, we know the cause and we have treatment. What more would massive research into TB tell us? The main problem for us with TB today is one of logistics—developing and providing drugs and improving nutrition. It is also the problem of HIV, because in many instances HIV has been responsible for making TB even more crippling. In fact, this is true with most infectious diseases: HIV makes them worse.

Another factor to consider is that just as AIDS research has benefited, especially in the early 1980s, from information derived from other areas of basic and applied medical research, so are other areas now benefiting from AIDS research. AIDS has also produced an epidemic of some cancers, such as some forms of B-cell lymphoma and Kaposi's sarcoma, and increased the incidence of a few others. New insights into the origin of these tumors has come out of AIDS re-

*Before this is overinterpreted I should emphasize that heterosexual HIV transmission can occur in both directions—man to woman or woman to man—even in the absence of any venereal disorders.

search, which I will discuss in chapter 15. It is very likely that this information will be helpful to our understanding of some other cancers as well. Vaccine research, research in basic virology, immunology, and molecular biology have already been greatly enhanced by AIDS research. So has antiviral chemotherapy. For these and other reasons, I conclude that the attention to and the funding of AIDS research has not been in excess. If anything, it has reached a reasonably adequate level a few years late.

8. *Will the AIDS virus go away?* It is doubtful that this will occur without successful intervention by medical science or a drastic reduction in the number of infected people due to changes in living habits. What is likely is some stabilization of the rate of infection of humans with HIV. This may lead to a chronic acceptance of HIV, as society comes to view it as a tolerable hazard. The outcome in this case may be analogous to the situation with syphilis prior to effective treatment. This would mean that generation after generation would accept the small (in most nations) but significant possibility of being infected with the AIDS virus.

9. *Since the AIDS virus frequently mutates, will it mutate into a virus that can be casually transmitted?* At the AIDS International Conference in Stockholm in June 1988, both Montagnier and I were asked this question in separate, simultaneous interviews. At that time, we gave different answers. He said it was possible. I said if it happened, it would probably mean the end of the human race, but I thought it was impossible. The ability to be casually transmitted would change the very essence of the virus. The reason for its lethal effect is, in my mind, chiefly due to its uncanny knack of targeting the most critical cells of the immune system. It does this by a very specific interaction of its envelope glycoprotein with the CD4 molecule of T4 helper lymphocytes and macrophages. Casual transmission would require the virus to be much more readily capable of infecting the cells that form the lining of the ports of entry, such as the nose, throat, mouth, or gastrointestinal tract—the mucosal epithelial cells. It is hard to conceive of HIV replicating sufficiently and concomitantly altering its cellular tropism in a manner that would allow it easily to infect these cells and still maintain its tropism for the T4 helper lymphocyte and the macrophage.

When Montagnier and I met in private and had a chance to discuss our views, we agreed to respond to the question with a middle position: the mutation of AIDS into a casually transmitted virus would be highly improbable. I think there are too many real problems to worry about with AIDS, other health issues, and the environment to consider seriously this kind of fantasy. Variations in the AIDS virus do occur, but in regions where the variation is tolerable. And the degree of that variation has limits.

10. *How much HIV-contaminated blood or saliva does it take to get infected?* Here is one of the most common questions of health care workers and others who can be exposed to people's blood and saliva—the nurse who accidentally pricks his or her finger with a needle, the surgeon who cuts his or her finger while operating, the person giving mouth-to-mouth resuscitation to an HIV-infected person, and so on. I have seen scientific estimates of the volume of blood necessary to transmit infection ranging from 10 microliters to 100 microliters (100 microliters is about 2 drops of fluid; 10 microliters is roughly equivalent to the amount needed to cover the head of a pin, or less).

To me these estimates do not have much meaning. Obviously, the answer to the question depends upon the amount of virus in the blood (or saliva). With the exception of a very transient period after the onset of infection (when sufficient time for an immune response has not yet elapsed), the amount of HIV in the blood usually is much higher late in the course of AIDS than early in the disease. Thus, blood from a person in the late stage is more likely to be infectious. As we will see in more detail in the next chapter, exposure to one virus particle is unlikely to bring about infection (many virus particles are defective, others may lose their envelope, still others may be degraded before they find their target cell). Of course, the nature of exposure to HIV-contaminated blood is also critical. Contact with unbroken skin probably does not ever lead to infection, but exposure to broken skin (for example, skin that has an inflammation, a dermatitis) could lead to infection. We have seen examples of this when the exposure was to a considerable amount of concentrated virus (a laboratory accident in a facility producing large amounts of the virus).

I do not know the current precise statistics regarding infection of health care workers from accidental sticks with needles contaminated by HIV-containing blood, but everyone agrees it is a rare occurrence.

Here the volume of blood exposure is very small. Therefore, unless the amount of virus present is high, transmission will not occur. There are case reports of blood exposure to the eyes and face leading to infection of health care workers. In these cases the volume of blood has been large, as in a laboratory accident.

The concept of saliva as a mode of transmission has been controversial. I can state only that my co-workers and I found and reported detectable HIV in the saliva of some infected people. The amount of virus was very low, but I do not believe we can dismiss saliva as a potential source of infection when there is substantial exposure. Reports from the Soviet Union of apparent transmission of HIV *from* infected babies (infected by exposure to contaminated blood) *to* their mothers by nursing is illustrative in this regard. Yet, we should not take this notion too far. For example, there is no evidence for HIV transmission by kissing, or, of course, by droplets in the air, as from sneezing or coughing.

11. *Can't we get drugs to sick people faster? Aren't there excessive delays?* Of course, the answer is and must always be yes. Yes, we can always do better at getting drugs where they are needed faster, not just to AIDS patients but to all those suffering with serious disease. But I cannot think of a more sharply two-edged sword: the faster drugs are moved into therapy, the greater the risk of incompetence, serious toxic side effects, even outright fraud. Numerous treatments for AIDS have raised instant headlines, caused harm, then disappeared. Those who sped them into clinical use by vocal support or "activism" then simply stop talking about them. No apology is given. Nothing is taken back. No correction in the "literature" in which the "success" was first announced (usually newspaper) is made.

The harm is not only to the health of the treated person. Much of the time such drugs have been physically harmless, though not always. Their ill effects have usually been measured in time lost by scientists and health care workers who were pushed—sometimes even forced—to chase rainbows. Harm can also come from fostering inappropriate loss of confidence in medical scientists and health officials or in a patient's own physician. These unwanted effects are only part of the problem. Not taking more appropriate health measures is the more serious corollary. The most recent example of a new "panacea" for AIDS is hyperthermia, heating blood to kill the virus. I hear many

HIV-infected people talking about this with more than a little interest. But the procedure has serious hazards associated with it, and makes little sense. In the end, despite some early scientific supportive voices, I imagine it will be seen as producing more harm than good.

Yet we know that sometimes rationally based approaches to treatment, accompanied by sound laboratory experimental data, can lead to the development of a drug for AIDS that then takes a very long time to enter the clinic. The reasons for such delays are not always simple and not always due to our inabilities or any excesses in FDA regulatory policies. Sometimes, for example, they are due mainly to lack of availability of the compound. I may have an idea, develop a laboratory system to test it, find a chemical compound based on the idea and experimental data, get positive laboratory results, have adequate toxicology data, but be frustrated in my ability to get the compound to the clinic because of my dependency on the manufacturer's interest, capabilities, and corporate decisions.

Fortunately, there are highly intelligent, well-informed activists who keep close watch on both inappropriate claims and unnecessarily slow movement of drugs. In this regard, Martin Delaney's Project Inform plays a pivotal role. With keen insight and lucid discussion, this organization constantly evaluates information on AIDS therapy, distills it, and distributes it to many thousands of HIV-infected people.

12. *Is there a relationship between AIDS and cancer?* Yes, and the relationships are several. First of all, many people with AIDS develop one or more cancers, as we will see in chapter 14. Sometimes these are apparently related to other infections. Other times they seem part of the very process of HIV infection. Second, AIDS is caused by a retrovirus, the third one known in humans. The first can cause cancer of T4-cells (adult T-cell leukemia), and most retroviruses of animals are known to cause cancer. Third, the long-term disease, the suffering, and the complications remind one of the experiences of many cancer patients, as does the social stigma in the early years of recognition of these diseases. It was, after all, not so long ago that people were afraid of the very word *cancer,* and almost seemed to hide or want to hide a relative with the disease.

Lastly, many principles of AIDS therapy use cancer treatment as a model: the overall goals are the same—selective targeting of a cell or a virus. Combination chemotherapy, drug resistance and how to avoid it, extensive medical and social supportive therapy, complicated clinical trials, and ethical questions related to clinical trials, were and still are part of everyday cancer treatment programs.

On the other hand, the fundamental cellular and molecular disturbances—the very nature of the disease—are, of course, substantially different.

13. *When will there be a cure for people with AIDS and a successful vaccine to protect people from being infected?* Before considering the answer to this question, we should first maximize our understanding of how the virus works and then consider the evidence that HIV is the cause of AIDS. These are the subjects of the following chapters.

13

How the AIDS Virus Works

The clinical manifestations or first symptoms and signs of AIDS seem to me to have five aspects, some of which are completely independent of the others.

1. *Serious abnormalities of the immune system with the inevitable consequence of opportunistic infections.* Decline in immune function will lead to an eventual inability to guard properly against many infections. Opportunistic infections are those with microbes that normally do not become established in a person and/or do not cause disease. But in certain situations, such as in a person with a dysfunctional immune system, they do. For reasons not understood, the most common serious opportunistic infection in AIDS is with a ubiquitous airborne parasite known as *Pneumocystis carinii.* It can be treated—though not always successfully—with an oral antibiotic, Septra, followed by a drug called pentamidine, which is given intravenously and can also be used, in an aerosol inhalant form, as a prophylactic against the infection.

A great variety of other opportunistic infections, such as skin lesions, also occur, owing mostly to other fungal, parasitic, or viral infections. A strange and not uncommon infection also occurs with a bacterium not usually associated with humans. Known as Mycobacterium Avium Intracellulare, this relative of the bacterium that causes TB normally infects only birds.

Although this agent is a bacterium, infections by bacteria are not as numerous and usually not as serious as infections with viruses and fungi. This seems to be because harm is chiefly to the cellular arm of the immune system, with much less damage to proper antibody production (humoral response), and it is the latter that chiefly protects us against bacterial diseases. This is less true in infant AIDS, however.

Many very common viral infections of humans that are usually innocuous can become serious in a person who has AIDS. One herpes virus, the cytomegalovirus (CMV), is a good example. Although many people are infected with CMV all their lives with few or no effects, people with AIDS often show serious symptoms from this infection. Perhaps the best known is a severe inflammation of the retina, the lining of the interior of the eye. Called CMV retinitis, this condition can lead to blindness from bleeding or retinal detachment.

Sometimes people with HIV infection develop anemias. Of the several ways this reduction in the number of red blood cells occurs, one is a failure to make enough of the red-blood-cell precursors in the bone marrow. Studied chiefly by Neil Young and his associates at NIH, a type of small, very common virus known as parvovirus, not harmful to healthy people, can produce severe aplastic anemia in HIV-infected individuals.

Another herpes virus (called herpes simplex) produces lesions of the skin and the membrane lining of the mouth, lips, and esophagus. Still another, Varicella-zoster, causes shingles—a terribly painful condition of the nerves and skin. Several fungal infections other than pneumocytis pneumonia occur with high frequency and also can be serious. One such disease that is sometimes similar to TB is known as histoplasmosis.

Another fungus (candida) causes thrush, a painful disorder that gives the mouth a white, patchy appearance. Cryptococcus is a fungus similar to candida. Many normal people carry it in their lungs, but the immune system "holds" it there. In AIDS patients, it may disseminate and reach the membranous outer layer covering the brain and spinal cord. If so, it can cause severe meningitis.

Finally, serious parasitic infections also occur. Some cause fever and diarrhea, others even more serious problems. Perhaps the best known is toxoplasma. This parasite is transmitted to humans from animals, infecting and being carried by a significant percentage of healthy people (10 to 30 percent, depending on the geographic region).

Usually it is localized to the muscle or brain. In AIDS patients, it can lead to serious destruction of brain tissue (toxoplasmosis of the brain).

These and many other opportunistic infections occur as the T4-cell count declines. The normal number of T4-cells in our blood is about 1,000 per cubic millimeter of blood. The T4-cell count drops about 50 cells per cubic millimeter of blood per year after HIV infection. When the number falls below 300, more and more serious infections occur.

Treatment for almost all of these infections is extremely difficult, and further research is vital. More than anything else, it is the myriad infections that result in the unusual diversity of the clinical findings in AIDS—and therefore it is called acquired immune deficiency *syndrome* rather than *disease.*

2. *Reaction of functional components of the immune system to the virus.* Despite serious specific impairments in immune function, infection with HIV does not destroy all immune function, and obviously there is only slight impairment in the early stages of infection. And despite the *need* for a potent immune response to HIV, sometimes the immune response itself causes symptoms and even serious side effects. So during early periods after infection some of the clinical signs and symptoms may be directly due to the immune response to the virus. For instance, there is often fever associated with viremia (free virus in the blood), which very likely is due to the release of temperature-elevating molecules derived from immune-stimulated cells. Enlargement of the lymph glands is also common in early stages, partly due to the multiplication of lymphocytes in response to the virus.

Blood platelets, or thrombocytes as they are sometimes called, are pieces of a cell that are formed by pinching-off of a parent cell in the bone marrow. They are essential for blood clotting. With a reduction in their number, easy bleeding occurs, manifested as easy bruising, blood in the urine or feces, nosebleeds, and the like. Platelets can be reduced after infection by antibodies made against an infecting microbe, which sometimes occurs following HIV infection and is known as *immune thrombocytopenia.* A phenomenon known as *lymphocytic interstitial pneumonitis,* relatively common in pediatric AIDS and one of its most serious and acute aspects, occurs when white blood cells—lymphocytes and macrophages—fill the lung cellular lining, interfering with the passage of critical oxygen from the lungs to the blood capillaries and of carbon dioxide from the bloodstream to the lung cellular lining and then to the outside. Essentially the infant or child is suffocated. Although the reason for this reaction is not certain, one suspects that HIV-infected lung macrophages release chemical attractants (chemotactic factors) which attract other cells to infiltrate. This too is an example of a disorder resulting from a normal immune response needed to control HIV that produces harmful side effects.

3. *Infection of the brain and resultant thinking disorders,* often progressing to dementia and/or psychosis.

4. *Induction of certain kinds of abnormal cell growth,* such as B-cell lymphoma and Kaposi's sarcoma.

5. *Miscellaneous disease due to an increased incidence of infection with disease-causing microbes (pathogens).* Pathogenic microbes differ from opportunistic ones by causing disease with some predictable incidence in healthy persons. Because of lifestyle—for example, the repeated exposure to a variety of specimens of human blood in the case of a drug addict—and because of the immune dysfunction in HIV-infected people, there is not only the increased incidence of opportunistic infections but also a much greater risk of infection with such microbes. This includes infection with some forms of hepatitis viruses (which can sometimes cause not only hepatitis but liver cancer as well), herpes simplex viruses, papilloma viruses, the bacterium of tuberculosis, the HTLVs, and many more.

Let us now go through what I understand to be the various stages of infection of an individual by HIV, using the example of transmission of HIV from a man to a woman during sexual intercourse. The infected man in our example has an increased number of white blood cells in his semen as a result of a past venereal disease, which substantially increases the amount of HIV because some of the macrophages and T4 lymphocytes contain HIV. (Though this may not be essential for male-to-female sexual transmission, I think it is obvious that it will increase the dose of HIV and thus the probability of transmission.) There are also a small number of free extracellular viral particles. When the man ejaculates, the few HIV particles may lose their outer coat (envelope) and most may not be infectious, but some infected T-cells and macrophages enter the woman's bloodstream by way of minor tears in the vaginal mucosa, perhaps only microscopic in size. In some cases, so may some of the infectious free virus particles.

The macrophages harbor intracellular infectious virus,* and many intact and infectious HIV particles are released as the macrophages degenerate in the foreign environment. These particles find waiting targets as they drift through the bloodstream and begin to migrate into some tissues. (Macrophages have a property, called *diapedesis,* by which they move themselves, or projections from themselves, into the spaces between cells.) Present in tissues throughout the body are Langerhans cells, which appear to be related to macrophages. Langerhans cells also have CD4 surface molecules. Consequently, they get infected. Probably more important, transmission of HIV occurs early after infection when the cell carrying the HIV (or, by now, the free virus) enters the lymphatic system.†

Here in the lymph nodes are follicular dendritic cells. These too are probably related to the macrophage, have CD4 molecules on their surface, and therefore are also targets for HIV. Like the blood macrophage and the Langerhans cell, these cells also process and present foreign products (antigens) to the lymphocytes. T4 lymphocytes, migrating into the portion of the lymph node known as the germinal center (because of the multiplication and development of B-cells in this region), interact with the follicular dendritic cells that are present in those centers. Numerous HIV particles, having replicated in the follicular dendritic cells, are ready to infect these T4-cells. This early period is often accompanied by lymph gland enlargement (lymphadenopathy) partly, I assume, because of the proliferation of B-cells to make antibodies against HIV.

These B lymphocytes are activated, that is, they are in an active metabolic state, often undergoing cell division. In addition to making and releasing antibodies, some B-cells will also release lymphokines, biologically active protein molecules I have already touched upon. Some such lymphokines promote more HIV infection, and the proteins of degraded HIV may in turn make more lymphokines by activating more B-cells. A not unrelated vicious cycle of this kind may

*This is an unusual phenomenon. I noted earlier that fully formed infectious retrovirus occurs only after the budding-off of an infected cell's outer membrane. For HIV-infected macrophages, the conventional mechanism occurs, but fully infectious HIV also occurs *within* the macrophage.

†This may happen in different ways. I think the major way is when a blood lymphocyte or free virus enters capillary terminals in the lymph nodes, known as sinusoids. Here there are intercellular openings through which cells and virus can enter the lymphatic system. The capillary terminals are the tiniest endings of the arteries carrying oxygenated blood from the heart to all tissues.

also occur with T4-cells, though with these cells, the induction of lymphokine release is probably not by HIV proteins but by HIV itself infecting these cells. Such phenomena may lead to vicious cycles of virus (or viral protein) promoting the manufacture of lymphokines; and this leads to more virus, and so on. In the early years, however, the immune system is still able to handle the challenge. Not so later.

But as Peter Biberfeld of Stockholm and others have shown, the lymph glands gradually degenerate. It seems that the follicular dendritic cells are ultimately destroyed by HIV. The "architecture" of the gland slowly loses its form, and it no longer provides an acceptable home for the lymphocytes to develop.

In another possible scenario, the infected T4 lymphocytes from the man's semen may also have entered the bloodstream, but most do not have intracellular infectious virus—only the "silent" provirus, the viral genes in the DNA form integrated into the chromosomal DNA of the cell. But, prior to its death, this infected T4 lymphocyte may meet head-on with one of the woman's cells and thus become "activated," a well-known phenomenon in experimental cell culture systems. It can also become activated by infection with other microbes or by the effects of certain lymphokines that may already be present in sufficient amounts in the blood due to the earliest effects of HIV and HIV proteins on other cells. If so, the DNA provirus genes will be turned on, and HIV will form, release, and be efficiently transferred to the recipient cell by a process I assume is similar to that seen in the laboratory: the most efficient means of transfer of any of the human retroviruses, co-cultivation, mixing infected cells with uninfected ones at cell concentrations designed to ensure cell-to-cell contact.

It is important to understand that one exposure is all it takes. Because many people escape infection from one or more exposures, some clinical scientists in past years were confused into believing that repeated exposure was necessary for infection to occur. But this makes little sense. The far more likely cause is *chance* exposure to a sufficient amount of virus during any one contact—and the *dose* of the virus is most important. Many particles are inactive, and only some will make it to the target cells in an infectious form. The greater the number of virus particles, the greater the probability of this occurring. The strain of virus may also be important, because laboratory experiments show that some variants replicate far more efficiently than others.

It will take the woman in our example several weeks, maybe longer, to make antibodies to the virus and several days to make a cellular immune response. During this period, then, the virus has its field day and replicates extensively, so much so that scientists can readily find free virus in the blood, the viremia mentioned in the previous chapter. (This period may be accompanied by fever and chills and other indicators of an acute reaction.) This is the stage when HIV is able to establish itself.

What are the steps in this process? First, HIV must get to its target cell. If there are sufficient numbers of infectious particles, by sheer chance some may reach target cells prior to becoming inactive, in which case a component of the envelope of HIV locks into the CD4 molecule present on the surface of its target cells. Cell-to-cell contact is a more efficient means of virus infection, because the virus is newly formed and hence more apt to be infectious. In addition to the HIV envelope locking onto the CD4 molecule, there may also be other interactions that temporarily bind cell to cell.

Soon after the virus is bound to the target cell CD4 molecule, a process occurs by which the core of the virus penetrates the cell. This process is poorly understood, but the core houses the viral RNA genome and reverse transcriptase, both protected by the core structural proteins. In the cytoplasm of the cell, the provirus DNA is made from the viral RNA by a process called reverse transcription, which I have discussed in earlier chapters with reference to the HTLVs and to some extent with HIV. This reaction is catalyzed by reverse transcriptase and, as noted earlier, essentially provides a DNA form of the viral genes formerly existing in an RNA form. The process is not infallible, and mistakes made during it may lead to an inactive, "dead" virus, but such mistakes may actually strengthen the virus by contributing to its viral diversity and hence maximizing its chances of survival in the alien, changing world of the newly infected person's body.

Note that the conversion from RNA to DNA occurs *only* in the cell, never outside it. This is most probably due to the presence in the cell of the "building blocks" of DNA, the deoxyribonucleotides (which are not present in the virus alone). The deoxyribonucleotides will join together, forming strong chemical bonds (called covalent bonds) in the order instructed by the order of similar nucleotides of the viral RNA, acting as the blueprint for the new DNA. We can think of the attachments of one nucleotide to another as a string of pearls. Each pearl (nucleotide) has a hook to attach to a preexisting pearl on the necklace and a second hook at its opposite pole to receive the next pearl. The difference is that there are four kinds of pearls, and the order in which they join together is predetermined by the order found in the template RNA. This order will determine the function of each of the viral genes and whether the virus will survive even this early stage.

After the first strand of DNA is formed, a second one is made. Now the first DNA, rather than the RNA, acts as the master guide. In fact, by now the original RNA has been destroyed by another mechanistic function of reverse transcriptase. The core with provirus DNA enters the nucleus (the details of how this occurs are not clear), and the provirus integrates into the host DNA at many sites, apparently selected almost at random.* This completes the first half of the virus replication cycle.

But there are some important uncertainties. I have mentioned, for example, the strange occurrence of complete infectious virus particles formed and present within the cytoplasm of some of the macrophages (and sometimes even in T4-cells), which I have not seen with infection of cells by other retroviruses. How this occurs is unknown. Is it possible that HIV can replicate in these cells *without* provirus integration into the nuclear chromosomal DNA?

Whether or not integration occurs, the second half of the HIV replication cycle must activate the DNA provirus genes so that viral RNA can be remade. From virus cell culture experiments we know certain things that do turn on the viral genes. We have already seen one way this occurs—with infected T4 helper lymphocytes, by contact between unrelated cells. More commonly, activation is induced by proteins of a foreign agent such as occurs in an infection. Some proteins serve as this antigenic stimulus to some T-cells; other proteins, to other T-cells.

There is evidence that other cells, such as macrophages, take part

*For the DNA provirus to integrate into the cell's chromosomal DNA, there must be both openings or breaks in the cell DNA and a means to tie together the ends of the viral DNA with the opened ends of the cell's DNA. This process depends upon other components of the virus (other special enzymes) as well as on some things provided by the cell. It appears that the cell must be in the process of mitosis (cell division) for this to occur, particularly in the stage of cell division when the cell's DNA is being made (duplicated) in preparation for formation of two daughter cells.

in this process by first presenting the foreign protein to the T4-cell, and by secreting some biologically active molecules called *cytokines* and in some cases interleukins (when secreted by lymphocytes they are also known as lymphokines, as I have discussed*). During the activation process, the metabolic state of the T-cell dramatically changes. Many formerly inactive cell genes are turned on. New proteins are made. The cell becomes bigger and it divides. It is during this period that the DNA provirus genes of HIV are also activated.

The expression of the viral genes is controlled by a regulatory element called the LTR. Coincidentally, some specific signals in this element of HIV are almost identical to sequence signals in human cellular DNA. In T-cells these are affected by cellular factors—protein in their chemical nature—that appear when a T-cell is so stimulated; and when so affected, a group of inactive cellular genes becomes active and new proteins made so that the T-cell can carry out its function as a mature cell. In the case of a T4 helper cell, this stimulation can result in the production and release of other lymphokine proteins of critical biological importance; for instance, to help B-cells make antibodies and several other proteins with functions that significantly influence the immune system. Unfortunately, the same stimuli will also activate HIV expression in an infected T-cell.

But before HIV can be formed, several steps are still needed. The fate of the newly synthesized viral RNA varies. Some will directly exit from the nucleus to the cytoplasm. Just as we have seen with the HTLVs, others are spliced (once or twice). This process is like its namesake in cinema in that a sequence (of RNA) is cut, some portion(s) removed, and the remaining parts ligated back together. Whether or not the RNA will be spliced and, if spliced, whether once or twice turns out to be critical to the final function of the RNA. Those not spliced, or spliced once, form the messengers that direct synthesis of the viral structural proteins. Those spliced twice form the messengers that direct formation of the regulatory proteins.

The entire process of viral RNA formation, movement, and splicing depends upon at least two viral proteins (called *Tat* and *Rev*), which we will call regulatory proteins. The formation of these viral regulatory proteins is, as I have said, directed by some of the doubly spliced

*One lymphokine made and released by activated T4-cells is gamma interferon. This substance can in turn activate macrophages so that when a macrophage "ingests" a foreign invading microbe, it can degrade it.

RNA, which functions as messenger RNA for their synthesis when this RNA gets to the cytoplasm. These are the first viral proteins made when an infected T-cell is immunologically stimulated. Therefore, they are also called "early " viral proteins and the genes that code for them, "early" viral genes. Later, unspliced RNA or singly spliced RNA molecules will function as messenger RNAs to direct synthesis of the viral structural proteins ("late" viral proteins). Some of the viral structural proteins (those that will become the core proteins) must be "cut" to smaller size. This is managed by a specific enzyme known as a *protease* (which functions quite differently in the AIDS virus than it does in normal cells).* Note the similarity of the overall scheme to the replication of the HTLVs.

This protease is specific to the AIDS virus. The final step is the assembly of viral genomic RNA with viral structural proteins at the cell membrane and virus formation by budding-off of the cell membrane. As I mentioned, HIV is peculiar in that some virus forms within the cell cytoplasm, at least in some infected cells. HIV has still other genes, and they make proteins whose functions are poorly understood, although it is possible that they, too, are involved in regulating the sequence of events that leads to the formation of HIV.†

HIV has added another process to the usual way a retrovirus forms. This process, however, is similar to the one evolved by the HTLVs. These viruses have a way to hide from the immune system (the "silent," unexpressed DNA form of the virus integrated into the DNA of the chromosomes of an infected cell). They complete their replication cycle only when these genes are activated, but it will be to the advantage of the virus to do so with some abundance and to do it

*There are many kinds of proteases and several forms are present in normal cells. The protease of the AIDS virus, however, has some specific peculiarities. Its major known function is to "cut" a larger protein called a precursor protein to smaller size, which will then become the final structural proteins of the HIV core.

†Most of the details of these events were discovered and worked out by the same individuals who contributed to the earliest molecular biological studies on human retroviruses, especially by my co-workers Flossie Wong-Staal and Sasha Arya; William Haseltine and his co-workers; Craig Rosen (now at Roche); Joe Sodrowski in Boston; and Simon Wain-Hobson of the Pasteur Institute. Important and innovative contributions to the early molecular biology of the HIVs, and more recently also with the HTLVs, have also been made by George Pavlakis and Barbara Felber of Frederick Cancer Research Facility, Maryland; Beatrice Hahn and George Shaw, formerly in our group and now in Alabama; Irvin Chen of UCLA; Paul Luciw at the University of California at Davis; James Mullins, formerly in Max Essex's department at Harvard and now at Stanford; Warner Greene and Bryan Cullen at Duke; S. Venkatesan, M. Martin, and K.T. Jeang at NIH; Lee Ratner, formerly with us and now at Washington University; Masa Hatanaka in Kyoto; John Brady of NCI; Matteus Peterlin at the University of California in San Francisco; and Mike Matthews at Cold Spring Harbor Labs.

quickly. HIV does both well, and like the HTLVs does so through the function of one of its regulatory proteins. When this "early" protein (*Rev*), which is essential for virus formation, is present, then and apparently only then can the messenger RNA molecules be made available for forming the structural proteins of the virus. But this occurs at the expense of itself! When *Rev* acts to favor function of these messengers for structural proteins, its own messenger RNA is lost. No more *Rev,* no more virus production. A rather neat control that allows a burst of virus formation and a quick end.

As Daniel Zagury and I have shown, when HIV is released from the T4-cell, the cell may die, and this seems correlated with the amount of virus released. We do not understand why, but it correlates with the amount of envelope protein of the virus and the amount of CD4 of the cell, a finding first made by Haseltine and his co-workers at Harvard. Virus released will spread to more sites. Anthony Fauci and his colleagues, along with an independent group in Japan, have discovered that the macrophage can also be activated to release virus in conjunction with certain cytokines, small proteins with biological activities, that are likely to be released during inflammatory processes. Unlike T4-cells, however, the macrophage does not die, which is why some scientists suspect that it acts as a reservoir of virus. HIV may, however, induce the inappropriate release of cytokines from macrophages,* which may cause some of the disease symptoms. In a vicious cycle, these cytokines can promote more HIV production.

This type of laboratory study strongly suggests that the rate of AIDS progression may be accelerated by anything that activates T4-cells and macrophages. Thus, chronic infections in general that activate T-cells may be more deleterious to AIDS victims than they would be otherwise. Since such infections routinely occur in people with AIDS, epidemiological data do not distinguish them as co-factors in the disease. However, some recent studies strongly suggest that HIV may be directly and more profoundly activated by specific infections. For example, Courtenay Bartholomew and his co-workers of the University of the West Indies in Port of Spain, Trinidad, in collaboration with

*Earlier in this chapter, I mentioned the same kind of phenomenon with B-cells and T4-cells, though because they are lymphocytes, the biologically active proteins were referred to as lymphokines instead of by the more generic name of cytokines. But it all means much the same thing. Another difference is that it is HIV proteins, not infectious virus itself, that leads to B-cell activation. B lymphocytes do not have the CD4 surface molecule and therefore are not usually infected by HIV.

William Blattner at NCI, have shown that the rate of progression to AIDS in HIV-infected people is accelerated by simultaneous infection with HTLV-1. Similar findings have been made by Stan Weiss among drug addicts in New Jersey infected with HTLV-1 (or HTLV-2) and HIV-1, by Nashiko Hattori and Kiyoshi Takatsuki in Japan, and by other groups studying various HIV-infected people. So at this time we can say that the HTLVs are *the only known specific co-factors* for AIDS, enhancing progression of the fundamental immune abnormalities; while many other microbes, which we all are routinely exposed to, may also contribute, they do not do so sufficiently and/or they are so common, that they are not statistically evident.

In chapters 6 and 7, on the HTLVs, I noted that an HTLV regulatory protein called *Tax* can be released from infected cells. When the released *Tax* reaches another T-cell, it can bind to its surface and activate that T-cell. If that T-cell carries HIV, this will also activate HIV expression. This could be the mechanism by which HTLV may enhance progression in AIDS patients. An alternative mechanism came from studies by my postdoctoral Fellow Paolo Lusso. He showed that an HTLV and an HIV can infect the same cell. This can lead to production of a virus that has envelopes of both, which can increase its types of target cells.

A decline in the number and function of T4-cells is one of the hallmarks of AIDS. In a savage irony, T4-cells and macrophages are two of our most critical protections *against* infections—and therefore are at the heart of the battle against HIV—yet these are the very same cells that the virus targets. But it is not only direct killing by HIV that depletes T4-cells. At any given moment only 1 of every 50 to 100 T4-cells in an HIV-infected person seems to show infection, and in our experience usually fewer than 1 in 10,000 cells expresses virus at most times.

Other, less direct mechanisms also appear to be operative. Some scientists suggest that *uninfected* T4-cells can be attacked by immune killer cells in HIV-infected people. One way this may occur has been suggested by Dani Bolognesi and his group, who have observed that viral envelopes from degrading virus can bind to the surface molecule, CD4, of uninfected T4 lymphocytes. HIV-infected patients have antibodies that react against this complex of envelope and CD4, and these antibodies attract killer T-cells by means of a nonspecific portion (called Fc) of the antibody molecule. Another possible co-factor in the depletion of T4-cells is via an abnormality in the production of the T-cell growth factor, IL-2, by T-cells of AIDS patients, even by cells not infected by HIV.* Still other mechanisms have been described by Ellis Reinherz and his group at Harvard. Finally, there is insufficient knowledge about the function of the thymus gland in HIV-infected people; if impaired, this gland, the seat of T-cell maturation, could lead to defective production of mature T-cells.

Back to my example of the recently transmitted infection: weeks later the virus has been disseminated to many sites of the body, chiefly by delivery from infected macrophage which can migrate between some intercellular junctions of small blood vessels to take residence at distant tissue sites, but probably also by dissemination of free HIV particles. Some of the brain microglial cells are now infected (in fact, some researchers believe these brain cells are directly descended from macrophages). Infected microglia may inappropriately release cytokines, and some of these cytokines can be toxic to nearby cells.

In our first examinations of postmortem specimens from people who had died with AIDS, we were very surprised at the high frequency of detection of HIV genetic information in the DNA of the brain. This was later confirmed in our laboratory and by others who reported frequent isolation of the virus or detection of antibodies to the virus in the cerebrospinal fluid, the liquid surrounding the brain and spinal cord. These results indicate that the brain is an important site of virus infection. Since some studies suggest that infection of the brain occurs early, direct effects of HIV in the brain are likely to account for the thinking disorders that can even be an early sign of HIV infection.

Prior to these findings, clinicians thought that most or all of the brain abnormalities were due to opportunistic infection with the parasite toxoplasma, or to the presence of a tumor (lymphoma) in the brain. Although these can occur, probably the vast majority of brain abnormalities in AIDS result from HIV infection. Of course, this finding had immediate practical importance because it changed approaches to treatment and diagnostic procedures of individuals with neurological abnormalities.

*In studies with Daniel Zagury, we came to the opinion that this phenomenon is due to the effects of another cytokine, alpha interferon, which is often present in high levels in HIV-infected people.

But what is the mechanism of brain damage by HIV? Some of it no doubt is a direct toxic effect of the virus on the microglial cells. This, in turn, may lead to less "architectural" support to the neurons, whose function may then be compromised. I have already alluded to another mechanism: the inappropriate release of cytokines. Volker Erfle suggested still another mechanism at my annual lab meeting in August 1990. He finds that another HIV regulatory protein (called *Nef*) is also released by infected cells and that *Nef* has some regions of similarity with a brain-poisoning protein from scorpions. Apparently, direct testing of *Nef* on various brain cells results in their developing abnormal cell membrane functions. These various mechanisms suggest that even with a relatively small number of cells of the brain infected, HIV may still be able to induce profound abnormalities.

Dissemination of infected macrophages to other body sites may account for many other relatively early disease manifestations. Another common sign is diarrhea, which can obviously be incited by many kinds of associated infections in an immune-suppressed person, but may also be due to direct infection of cells within the bowel wall by HIV, as suggested by a few studies.

In most infected people the immune response to the virus will include antibodies that can inhibit further virus infection (neutralizing antibodies) as well as cellular immune responses, such as the activation of killer T-cells that attack virus-infected cells. This tends to engender a false sense of security once the clinical situation appears to stabilize, whereas the virus may be slowly but inexorably diminishing the ability of the immune system to function during this period.

Just as bad, small genetic changes in the virus occur, generating mutant variants. Many, perhaps even most, of these variants would have been present at the beginning of the infection, but some of them would take time to emerge. Inevitably, those variants dominate that best avoid the attacks by the immune system, and so the process monotonously repeats itself. If the virus can gain a significant advantage over the immune system, then a whole series of harmful effects will follow: more virus, more infected cells, more harm to the immune system. It usually does.

Some clinicians speak about latency in AIDS. They imply that after the original infection, viremia, and acute response, there is a long period of viral inactivity after which the virus again becomes active and leads to the final stages of the disease. But this is not the case. We can think of latency only at a cellular level—that is, many cells are not expressing virus. But there are always some cells that do. Disease progression is a consequence of the accumulation of stimulatory factors, more virus spread, more immune suppression—until a threshold for the immune system is reached, below which are fatal consequences.

I deliberately chose sexual transmission from man to woman for my example of HIV infection in order to emphasize that this is probably the most common pathway of infection in the world—*not* man to man by anal sex, as we in the United States tend to emphasize. Because HIV infection showed up full force first among gay males in the United States and in Europe, it has been discussed far more in those regions.

The accumulating evidence, however, dictates that globally the number-one route of transmission is from man to woman. In man-to-man transmission, the virus enters from the semen into the bloodstream through minor anal or rectal tears. Had I chosen that example, I would have noted that my new co-worker Anita Agarwal, working with Veffa Franchini, discovered that two parasites found in fecal material, giardia and amoeba, can be infected by HIV! These parasites are not uncommon in central Africa and in gay men everywhere. We wonder whether this is a factor in HIV communicability.

The development of AIDS in a baby, of course, follows a different path. First, infection occurs chiefly within the womb by movement of free virus from the mother across the placenta and into the fetal circulation. For reasons we do not understand, some infants escape infection during pregnancy but may become infected from blood contamination during delivery, or even later from mother's milk. It appears that more HIV replication occurs, more cells are infected, and the course of disease is far more rapid in infected infants than in infected adults. These events are due to the newborn's less developed immune system, giving it even less chance to fight HIV. Other clinical differences between infant and adult AIDS reflect the different microbial environments as well as different degrees of tissue maturation.

Increasingly I am of the opinion that when HIV infection finally reaches a stage where immune function is measurably depressed, then other viruses may enter the picture, make matters much worse, and contribute to the progression of the disease until it reaches the full clinical abnormalities that physicians at CDC have coined AIDS. As I mentioned, perhaps the best example of another contributing virus is HTLV-1. I have remarked that it is difficult to provide epidemiological statistics for other microbes because they are so common and their effects significant perhaps only fairly late in disease so as to be undiscernible. But in a few instances laboratory experimental studies with such agents are suggestive of a possible contribution to progression: one that might contribute to the *primary* disease process of immune dysfunction, particularly the critical T4 depletion, is a new human herpes virus.

As I have already briefly discussed, herpes viruses are DNA viruses that contain far more genetic information than a retrovirus. Consequently, these viruses have many more proteins and are much larger than retroviruses. Herpes virus infections can cause an acute disease, usually self-limited, upon first exposure (usually in childhood). Like retroviruses they can also persist for the life of the individual, but they do not integrate their DNA into the host cell DNA. Rather, they often remain hidden in some reservoir cell population, with little or no replication until some stressful situation or immune suppression alters this balance with the cell. Then their active replication produces cell damage. Sometimes it is the immune response to them that brings trouble. There are five known human herpes viruses: herpes simplex 1 and 2 (the causes of cold sores and genital lesions); herpes varicella-zoster (the cause of chicken pox and shingles); cytomegalovirus (which can cause various problems in immune-suppressed people—particularly notable is a severe inflammation of the retina of the eye); and Epstein-Barr virus, or EBV (the cause of infectious mononucleosis and a contributing factor to certain human cancers, including some B-cell lymphomas and cancer of the nasal pharynx).

In 1985 my co-workers and I were trying to find the cause of the B-cell lymphomas that occur with extreme frequency in HIV-infected people. Lymphomas are cancers that occur when a lymphocyte in a lymph node or elsewhere becomes transformed to a malignant cell and multiplies. In a process called *metastasis,* the cells usually spread to other sites, especially other lymph nodes and the spleen. These organs become enlarged; some may become obstructed and/or otherwise damaged. Sometimes the cells also multiply excessively in the intestinal wall, in the brain, or in other sites. Oddly, the lymphoma in AIDS does not involve the lymph nodes much. More characteristically, it involves the walls of the gastrointestinal tract, leading to site-specific signs and symptoms such as pain and difficult swallowing when in the esophagus, or stomach and bowel problems when lower.

B-cell lymphomas are, of course, lymphomas that involve the transformation of a B-lymphocyte. Many of the B-cell lymphomas in AIDS patients greatly resemble African Burkitt's lymphoma, which, as I have discussed, is apparently caused by a combination of chronic stimulation of B-cell proliferation by chronic malaria infection and the herpes virus EBV, which allows many B-cells to be immortalized, especially when EBV replication is not controlled. But many of these AIDS lymphomas are EBV-negative. That is why we looked for another virus. My former co-worker Zaki Salahuddin, a collaborator, Dharam Ablashi, and I found such a virus the same year we began to look, and reported it the following year, 1986. We called it HBLV, for human B-lymphotropic virus. It was the first new human herpes virus that had been discovered for twenty-five years. Our results showed that the virus was widespread and was probably a very old infection of humans. Some data suggested to us that it might be a causative factor in some of these lymphomas. Fortunately, we did not promote this idea because now it seems likely that it is wrong. We were misled by the early isolation of the virus from B-cell lymphomas.

In another of the ironic and unpredictable twists in science, shortly after joining my laboratory a young postdoctoral Fellow from Turin, Italy, Paolo Lusso, convincingly showed that this virus chiefly infects T-cells. In fact, it principally targets the T4 lymphocyte! To my knowledge there were no viruses known to be chiefly T-lymphotropic—certainly none were T4-lymphotropic—until the discovery of human retroviruses. Now five of them had been found within the decade of the 1980s: the four retroviruses—HTLV-1, HTLV-2, HIV-1, and HIV-2—and this "new" herpes virus, which we now call human herpes virus-6, or HHV-6 (a neutral, safe name, just in case we received more surprises in the future).*

But it was not just that HHV-6 so efficiently infected the $CD4^+$ T4 helper lymphocyte that made it of interest. Lusso showed it killed these cells even more efficiently than HIV when it actively replicated

*In 1990, Niza Frenkel and her co-workers at NIH discovered another new herpes virus. They named it HHV-7—and it, too, targets the $CD4^+$ T-cell! Still more herpes viruses are being studied in my laboratory through the work of a new postdoctoral from Belgium, Zwi Berenman.

in them. Double infection with HIV plus HHV-6 gave a synergistic effect, killing more T4-cells than the additive effect of each of these viruses alone. Yet, I have said that HHV-6 is a worldwide, common, and almost surely old infection of humans. In the United States about 60 percent of children have antibodies and are assumed to be infected by the first year or two of life. There is evidence that in early childhood HHV-6 is the cause of a not very serious disease known as roseola infantum. Since herpes infections are usually latent—that is, they remain dormant for long periods, even for life—we assume that most adults have HHV-6 in some of their cells.

Why, then, don't most people develop T4 depletion and immune deficiency from this virus? I think it is because in a healthy person, herpes viruses are generally well controlled. When HIV infection progresses to produce some significant immune suppression, it is reasonable to believe HHV-6 replication may then advance. Our experience is in accordance with this idea because we have been able to find and isolate HHV-6 more easily in HHV-6–infected people who have AIDS than in healthy people infected with HHV-6.

Added to all of this is a third finding. Lusso recently found that HHV-6 can turn on the gene for CD4* in some cells that have previously been CD4-negative. For instance, HHV-6 infection of T8-cells converts them into cells with both T4 and T8 markers. Such cells now can be targets for HIV infection, because, as we have seen, the receptor for HIV on the surface of cells is the CD4 molecule. For these reasons I suspect that HHV-6 is also likely to be a contributing factor to the development of AIDS—*not necessary,* but perhaps speeding the progressive immune impairment.

But unlike the situation with HTLV-1, it will probably not be possible to obtain epidemiological data to support or refute this suggestion, which is solely based on *in vitro* experiments. This is due to the fact that most people are HHV-6-antibody-positive. Evidence for it might come out of animal studies, however. We have found that chimps are susceptible to HHV-6. We would like to know now whether the infection of chimps with HIV-1, which so far has not produced AIDS, will do so with HHV-6. Better yet would be development of a safe drug that specifically inhibits HHV-6 replication.

*The earliest and most popular name is T4 lymphocyte or T4-cell, but in recent years immunologists prefer CD4 positive T-cell.

This could be used in treating HIV-infected people in combination with anti-HIV drugs to determine whether doubly treated people do better than those on only anti-HIV treatment.*

Some reporters have asked me why I study HHV-6. Even if it promotes HIV and even if it kills T4-cells, what can I really do about it? To most scientists this is a strange question—we do not study only things from which we are *sure* something practical will emerge. Most of us study what interests us, and we study what is able to be studied. We follow where our leads take us. Very often it is just such approaches that lead to scientific openings. But the question is particularly inappropriate for work on HHV-6, because there is potential for a practical positive effect from research in this virus. If HHV-6 is sometimes a co-factor involved in AIDS progression, as I have speculated, it is likely that—even if I cannot imagine an approach at this time—someday someone else will learn how to interfere with it. There are already suggestions from clinical studies that a compound called *foscarnet* may help slow progression of AIDS; and my colleague Howard Streicher has shown that this compound inhibits an enzyme of HHV-6, thereby reducing its capacity to infect cells. It is possible that the two findings are connected.

As a parenthetical note on HHV-6, I should say something about chronic fatigue syndrome, also known as the "yuppie disease." Like EBV and a few other viruses, HHV-6 has sometimes been suggested as the cause of the mysterious syndrome of chronic fatigue described by, and found for the most part in, upper-middle-class people. There are some results compatible with this suggestion. Others have recently suggested a role for an HTLV-2–related retrovirus in these disorders. However, the evidence for both is, as yet, inconclusive.

Many HIV-infected people will develop one or more abnormalities of cell growth, both malignant and nonmalignant. I have heard some of my colleagues quickly summarize the origin of tumors in AIDS as the consequence of immune deficiency, as if that were an adequate explanation and we could go on to the next problem. This concept goes back to an old theory called immune surveillance, popularized many years ago by one of the most innovative biologists of our time,

*But finally, I must state that in this whole discussion I am theorizing about HHV-6 (and concluding about HTLVs) only as *contributory* factors. HIV is the only real cause of AIDS. Remove it, and the epidemic will disappear.

the late Sir Frank Macfarlane Burnet of Melbourne, Australia, Nobel Prize winner in 1960 with Sir Peter Medawar for the demonstration of immune tolerance. Burnet was also the originator of the clonal selection theory in immunology, which brought to light one major mechanism for the generation of immune diversity: the great range of immune responses.

This theory of immune surveillance is far less certain, and I think has often misled cancer researchers. The idea assumes that cells may often become neoplastic but are destroyed or kept in check by some components of our immune system. Some clinical observations correlate with this idea, but there are also problems with it. For instance, only a few kinds of cancer occur with increased frequency after prolonged immune suppression. Also, only a few forms of immune suppression increase the risk of any cancer. Hairless ("nude") mice are born with a defect in T-cell production, and they do not have increased tumor development. Many other examples illustrate the serious limitations in this concept. I will return to this in the next chapter.

I think that when we do see cancers or hyperplasias* in association with immune dysfunction, the mechanisms accounting for the abnormal growth are more complex. I have suggested that they are more likely to be from a positive growth stimulation from cells of the immune system than from the lack of function of cells of the immune system, which allow previously transformed cells to grow out of control. This seems to be more compatible with the restricted forms of immune impairment that lead to increased risks of tumor formation, and it is more in accordance with my own laboratory experiences. I am confident that we cannot generalize too much. Each form of abnormal growth has its own specific mechanism for its birth and growth.

We have our best illustrations of this in HIV-infected people. Several forms of abnormal excess cell growth have been described in people with AIDS. I have already discussed the lymphomas of B-cells. Another is Kaposi's sarcoma, which was so rare prior to AIDS that it has become almost part of the definition of AIDS. Some other forms of cancer, such as liver carcinomas, have also been described with

*The term *hyperplasia* indicates an abnormal growth of cells, but the cells are not cancer cells. They can be reversed to normal. Their abnormal growth is often temporary, often dependent on signals such as growth factors from other cells, and never metastatic. One can think of normal wound healing—if continued beyond the level of repairing, the wound is a hyperplasia. Malignant cells, by contrast, are inherently abnormal. Their signals to grow are abnormal and come from within.

increased frequency. I will consider now mechanisms possibly involved in the origin of each of these examples, beginning with the last.

There are no previous indications, to my knowledge, that immune impairment leads to liver cancer. Certain chemicals, like the peanut fungus–derived aflatoxin, may induce this malignancy. More important are some of the viruses that also cause inflammation of the liver (hepatitis) with varying severity. Because of this earlier known feature, we call these hepatitis viruses. Today we know there is more than one kind of hepatitis virus, and even though we now have learned that different forms belong to very different categories of viruses, they are oddly still named as though they all belonged to one group: hepatitis A, hepatitis B, and so on—a convenience that must be confusing for a new student.

Not all of these hepatitis viruses cause liver cancer. One that does is human hepatitis B virus, or HBV. Like human retroviruses, HBV is transmitted by blood, sex, and from mother to child. Thus, it "runs with HIV." Naturally, HIV-infected people have a higher incidence of HBV, and consequently of liver cancer, than do HIV-negative people. It is also likely that the deterioration of the immune system will allow for greater replication and spread of HBV, further increasing the probability of an HBV-induced carcinogenic effect.

I have already defined and described B-cell lymphomas and how I thought their development would be explained by mechanisms analogous to the B-cell lymphoma, known as Burkitt's lymphoma, endemic to parts of Africa. To briefly reiterate: a small number of B lymphocytes are infected by EBV, which immortalizes these cells. This occurs in most people. But the number of such cells is increased when there is a decrease in function of T8-cells, a subclass of T lymphocytes that includes cells that can govern virus expression and some aspects of cell behavior. In Burkitt's lymphoma, the culprit is probably the parasite (*plasmodia* is its formal generic name) of malaria.

B and T lymphocytes are unique among adult cells because the immature progenitors of the mature forms of these cells have evolved a mechanism for shuffling DNA sequences about the chromosome in a manner designed to create the diversity of sequences needed in their respective immune responses. For the B-cell, it is the making of the immunoglobulin, the antibody. This occurs throughout our lifetime. Such genetic manipulations surely must increase the probability of a mistake that may lead to a malignancy (the lymphoma), and this would be further enhanced by the increase in the number of such cells that do not properly die because of the effects of EBV. This is further aggravated by chronic stimulation to these cells to divide. The mitotic stimulus to a B-cell is a particular antigen, in this case the protein antigens of the malarial parasite.

I thought the situation would be parallel in HIV-infected people. The earlier T4 abnormality induced by HIV should lead to T8 dysfunction because of signals from T4-cells to T8-cells, and the T8 dysfunction in turn should lead to an increased number of EBV-immortalized cells. Greater mitotic stimuli to these cells from HIV proteins should also occur. But Riccardo Dalla Favera of Columbia University, a postdoctoral Fellow with me many years ago and now (and then too) a leading scientist in the study of oncogenes, showed that in about one-third of AIDS B-cell lymphomas, EBV could not be found. So we searched for other immortalizing viruses, leading us to the discovery of HHV-6 which, although a new herpes virus, does not immortalize B-cells.

We are left wondering about these EBV-negative lymphomas. If, as we presume, the cells must be immortalized to be a true malignancy, can it be that the amount and chronicity of exposure to HIV proteins is sufficient not only to do the expected—promote cell division of select B-cells by antigenic stimuli—but to do it so excessively that genetic mistakes are frequent enough to produce sufficient genetic change in some cells to immortalize them? Recent results from a San Francisco group and from Dean Mann at NCI suggest that this might be the case. Or is there another, as yet undiscovered, B lymphotropic virus that, like EBV, can immortalize these cells?

In any case, B-cell lymphoma is now occurring as a minor epidemic of a cancer due to HIV infection. Dr. Sam Broder, NCI director, has recently said that as many as 46 percent of people surviving longer with AIDS (on AZT) develop B-cell lymphomas. In the absence of AZT, with shorter survival, the incidence of this lymphoma is still extraordinary. One can predict major problems in the future for our cancer centers with patient care and cost.

Peter Biberfeld and his colleagues in Stockholm have recently found that one strain of the monkey virus SIV, which is the closest virus to HIV in the animal world, also produces B-cell lymphomas in monkeys called cynomolgus with exceptional efficiency (more than one-third of infected animals) and speed. These HIV and SIV B-cell lymphoma results are extraordinary and unexpected. Since neither HIV nor SIV infects B-cells (and indeed no significant viral sequences are found in the tumor), this means that the role of these viruses is unlike that of a conventional tumor virus. The infected cells are T4-cells and macrophages, yet somehow the cancer occurs in the uninfected B-cell. These viruses, then, do not transform a normal cell into a tumor cell; they work indirectly. This will be important to my next consideration, a strange tumor called Kaposi's sarcoma.

14

Kaposi's Sarcoma: A Special Tumor of AIDS

Since cancers usually occur within one or more of one's internal organs, they spread invisibly, sparing the patient and others the additional psychological horror of having to watch them grow. Kaposi's sarcoma, however, often occurs on the skin surface or in the mouth or other visible body openings. First described in 1872 by the Austrian dermatologist Moricz Kaposi, Kaposi's sarcoma remained a rare tumor, unknown even to some physicians, until the recent full-force impact of AIDS. Kaposi's sarcoma, or KS, as I will refer to it, was first seen in older men of Jewish and Italian ancestry, and later in others of Mediterranean ancestry, especially Greeks. It tended to be limited to the lower extremities and usually was not aggressive.

Much later it was found in central Africa, again usually in men, but this time in both young and old; in the young men who were afflicted, KS often involved more body tissue and was more aggressive than it was in older men. The Mediterranean form is now called *classical* KS, and the African form is often called *endemic* KS.

The cause of these tumors has been ascribed to genetic mechanisms, and some results support the possibility of inheriting susceptibility to the classical form. The African form is much more common in blacks and among those living in equatorial regions. Its preponderance in blacks (and in South Africa, among the Bantu tribe) is also consistent with a genetic factor. But the geographic clustering of the African form is, of course, also consistent with the possibility of an environmental factor. In fact, one virus—cytomegalovirus, or CMV—has been repeatedly studied as a possible causative factor, particularly by Gaetano Giraldo in Naples. I have already mentioned CMV as a common human herpes virus, sometimes causing serious problems in immune-suppressed people and in infants, but without any real harmful effects for most infected people. But the results of studies of CMV in KS are inconsistent and far from conclusive. In fact, in my view, the bulk of the current evidence argues strongly against CMV as a cause of any form of KS and it can be conclusively stated that if CMV is involved at all, there must be another essential, as yet unidentified, factor with it.

A third form of KS has been found in some patients who have been deliberately immune-suppressed in preparation for organ transplantation. It appears here, too, that Jewish and Italian ancestry is more common. The last known form is the new HIV-associated Kaposi's sarcoma, which is called *epidemic* KS and is also found predominantly in men—at least in Europe and the United States. This interpretation, however, is biased by the much greater HIV infection of male homosexuals in these areas but HIV-associated KS in Africa is also unusual in women. Like the endemic (non-HIV) African form, HIV-associated epidemic KS often occurs in the young. Its clinical course can be aggressive. Because HIV is quite common in parts of central Africa, the HIV-associated epidemic KS and endemic KS are often confused with each other. Obviously, KS has developed, and will continue to develop, in the absence of HIV infection. But this does not mean that HIV infection does not play a role (and, in my mind, the critical role) in the *epidemic* form. To me it only means that there is more than one way to increase the probability of developing KS, just as there is for a number of other cancers.

But KS is an odd "cancer." I cannot think of any tumor with such a preponderance among males, except in connection with the reproductive tissues or an *obvious* male-skewed epidemiological factor, as was once the case for cigarette smoking and lung cancer.

The very makeup of the tumor is also unusual. Most human malignancies are composed of a highly predominant cell type. A given type of cancer is usually spread by a monotonous invasion of a particular cancer cell, with allowances for some infiltrating normal cells. Kaposi's sarcoma differs. It is a mixture of cell types. Though predominantly a proliferation of small blood vessels (giving it a purple color, visible in skin lesions), infiltrating lymphocytes, and macrophages, also present are fibroblasts* and endothelial cells (the cells that line the inner surfaces of all blood vessels). In the midst of these, and becoming more dominant in time, are spindle cells, so named because of their shape.

Until very recently no one had been able to study spindle cells in pure form in the laboratory, but they are believed to be the true tumor cell of KS, with the rest of the mass composed of cells somehow responding to them. Most students of KS think that the spindle cells are derived from either endothelial cells or the smooth muscle cells that form the walls of blood vessels. But there is a problem with that view: different types of cells can take a spindle shape under some circumstances, and, as I have said, no one had been able to study KS cells in laboratory cell culture until very recently. So the precise nature and origin of these cells was not known.

Cancer, as we have seen, is clonal—the tumor cells are derived from one original transformed cell, and spreading occurs because some cells acquire the capacity to move to another site and grow there (metastasis). But KS seems to begin in many places at the same time. In the case of AIDS-associated KS, the lesions that may be found in the internal organs, mucosal surfaces of the gastrointestinal system, and skin, all appear to develop *de novo* rather than by metastasis from an original tumor site. If proved, this will be the oddest of all the characteristics of KS, implying that there is something continually present in an individual's internal environment that leads to development of abnormal cell growth at multiple sites over a prolonged period. Sometimes a KS tumor will disappear early in its development. For instance, patients who have been immune-suppressed in an attempt to avoid rejection of a foreign organ will often, as I said, develop the third form of KS. But when the immunosuppressive therapy is stopped, the KS tumor sometimes vanishes. *No* cancer behaves like this.

What is the current thinking, then, about the origin of KS and about the mechanisms involved in its development? Some of its unique attributes have led a few scientists to take the position, with which I

**Fibroblasts* are cells that make connective tissue. They normally grow when there is a wound. Their growth is normally controlled by external signals (growth factors) of more than one kind.

concur, that at least in its early phase, KS is not really a cancer but is due to some kind of chronic stimulation of cells by some growth factors. This reasonable idea is, however, vague and lacks data to support it. Usually arguments for it are linked to some equally vague notion about immune suppression, because of the *apparent* association of KS with immune suppression. Indeed, in most overview lectures on AIDS that I have listened to, if the AIDS-associated KS is mentioned at all, it is almost dismissed as due to the obvious immune suppression associated with HIV infection.

In part, this sometimes confident view derives from the occasional development of KS after organ transplantation and the immune therapy used to allow the graft to take, and from the fact that the epidemic KS is associated with HIV, a virus that ultimately seriously harms the immune system. But this view owes less to serious thought or serious investigation of KS than to the legacy of the late Sir Frank Macfarlane Burnet, whom I touched on in the last chapter.

Burnet was one of the most original thinkers in biomedical science. As director of the famous Walter and Eliza Hall Institute of Experimental Medicine in Melbourne, he pioneered many ideas important to modern immunology. Among his concepts is that of immune surveillance, which I shy away from because it has led to an overconfidence that we know the origin of a tumor in immune-suppressed people. In its simplest form, the hypothesis of immune surveillance is that cancer cells often occur but our normal immune system regularly attacks and destroys them. Hurt the immune system and you help cancer develop.

Although decline in some specific kinds of immune function may contribute to tumor progression, I doubt that immune suppression contributes much to the *origin* of human cancer. This does not mean that cells of the immune system are not involved in the origin of several cancers. But as I will soon discuss, my belief is that where the immune system is involved, its role may be not in its failure to act against an already developing cancer but rather in the very origin of the cancer or tumor—through an abnormal, inappropriate secretion of lymphokines and related biologically active molecules. Indeed, it appears to me that the concept of such aberrant release of these potent substances at the wrong time and in the wrong place will soon be of increasing importance to our ideas about the development of many diseases.

In challenging the concept that immune suppression itself is the key to the origin of KS, we can begin by asking, Why this particular tumor? Why not cancer of the lung, the skin, the pancreas, or some other organ? And what of classical KS and endemic African KS? I know of no evidence that immune suppression is regularly present in these forms. More to the point of our focus on the HIV-associated KS, Bijan Safai of Memorial Sloan-Kettering Hospital in New York, a dermatologist and longtime student of KS, was one of the first to study AIDS-associated KS. He has told me on several occasions that he has seen cases of KS as the first clinical sign of HIV infection. In other words, KS may occur soon after HIV infection, before any detectable impairment of immunity. A hardened believer in the concept of immune surveillance has argued that these KS cases, which develop after HIV infection but prior to detectable decline in immune function, caused changes in immune function that were simply too subtle for scientists to detect. This kind of idea or argument is, of course, not answerable. It makes for unending debates—and unproductive ones at that.

In my mind, despite the commonality of the pathologic lesions, the four different forms of KS are really four different diseases, having in common only their cell composition, their pathological-morphological description, and their predominance in men. In other words, I consider KS much like pneumonia or, for example, leukemia, for which the structural form of the lesion or the leukemic cells might look the same, but for several of which we can now distinguish different primary causes and pathogenetic mechanisms.

I have been asked how I can claim that HIV is new in man as an epidemic disease if Kaposi himself lived in the last century. This kind of question misses the central point. The classical KS (as well as endemic KS) is HIV-negative, just as tubercular pneumonia is negative for staphylococcus and all other primary causes of pneumonia, and staphylococcal pneumonia and pneumonias caused by other agents are negative for the bacterium of tuberculosis. But there is another point of interest here, a point that first got me interested in studying the cause and the pathogenesis of HIV-associated KS. Could there be another microbe, another virus, working conjointly with HIV, *necessary* for epidemic KS but not yet identified?

In 1985 Peter Biberfeld joined my laboratory as a visiting scientist.

He had been a pathologist at the Karolinska Institute in Stockholm for many years. With his longtime interest in lymphoma and in classical KS, he naturally fell into studies of the pathology of AIDS. When he joined me he talked about KS more than he talked about anything else. He had a vague, but generally I think correct, notion about the early KS tumor as being a non-neoplastic state. A few other pathologists, such as Alan Rabson and José Costa of NCI, had earlier proposed and published similar notions. Somehow a group of cells were being chronically stimulated to grow. We were impressed by the HIV link, but I knew HIV would not be in the tumor cells (presumably the spindle cells) because our knowledge about HIV suggested to us that it was not likely to infect these cells. If it did, there were many reasons that it was not likely to be this virus that directly induced their conversion into tumor cells. But we tested this possibility anyway, in order to be sure.

We probed the DNA of KS specimens for nucleic acid sequences of HIV by the technique of molecular hybridization. As expected, the results were negative. Biberfeld was little interested in speculating about what had triggered the whole phenomenon; rather, he focused on the mechanisms of its subsequent development. I felt strongly that HIV must be involved in this particular form—but how? And what about the other, earlier known forms? About this time I learned from epidemiologists that among HIV-infected people, homosexuals developed KS more than others. Later I heard that HIV-associated KS is on the decline relative to its rate a few years ago, though sometimes I hear from other epidemiologists that this point is debatable, that it derives from an inadequate and premature statistical analysis. Obviously, these findings suggested (but did not prove) that something in addition to HIV, even in the HIV-linked KS, was involved in the origin of KS. Could it be another accompanying microbe? Or was it some less specific co-factor(s)? The combination of these findings with Biberfeld's interest led us into more serious studies.

My first objective was to rule out the presence of a *known* candidate human tumor virus—that is, not HIV but a virus known in the past to alter cell growth. In a setting of HIV infection, it was possible that KS was caused by certain viruses, either known to be involved in other human cancers or known at least to transform human cells into cancer-like cells in a laboratory tissue culture setting. Our approach was the same one we had used to check for HIV. First, Edward Gelman, then a postdoctoral student with me and now the associate director of the Lombardi Cancer Center at Georgetown University, and later Stephen Josephs and others in my laboratory obtained KS specimens, purified the DNA from them, and tested them for the presence of the genes (nucleotide sequences) of a variety of viruses including HTLV-1, HTLV-2, HHV-6, CMV, papilloma viruses, and another DNA virus called JC.* Some of these experiments were performed with the collaboration of Peter Howley at the National Cancer Institute. The results were negative.

I then decided that we would attempt to grow the KS spindle cells in long-term culture. This would satisfy two objectives. First, it would offer the most likely pathway to find a *new* virus, one that we could not probe for in the DNA of the KS tissue because it was the hypothetical, yet to be isolated, KS virus—one that, of course, we could not have probes to detect until the first such virus had been isolated. Second, if we succeeded in growing the spindle cells, we could study their characteristics in pure form for the first time. Perhaps we would get some clues to the nature of these cells. Even if a new virus were not found, we might be able to gain some fundamental information about this tumor. The second objective was close to Biberfeld's heart. My former associate Zaki Salahuddin set to work with another visiting scientist, Shuji Nakamura, who had joined my group shortly before, coming from Professor Hideo Hiyashi's department of pathology of Kumamoto University in Kyushu, Japan. Nakamura and Salahuddin would attempt to grow the spindle cells, and Biberfeld would help in their characterization.

Long-term culture of the *right* cells in this mixed population would be necessary if we were to detect, isolate, and characterize a virus present in small amounts or in a quiescent state, but it would not be easy. We soon learned that even short-term culture of these cells had not been achieved. We began by trying various, by then well known and commercially available, growth factors. Most had little or no effect. A few gave us temporary growth of cells from KS specimens, but this was insufficient, not only for virus detection work but even for any detailed studies of these cells.

*JC viruses are small DNA viruses that show great similarities to animal viruses of the category papovaviruses, especially of its members SV-40 and polyoma viruses; these two produce cancer in animals, but there is no evidence that JC viruses cause cancer in humans. There is, however, evidence that JC virus can cause a serious neurological disease under certain circumstances, but not usually in healthy people.

We then decided to look for new growth factors. After some negative results we luckily found one that worked. Like our old T-cell growth factor (Interleukin-2), and our even earlier experiences with the factor for myeloid cells known as GM-CSF, and the many lymphokines since, the "new"* one is also released into the nutrient medium ("broth") by cultured T4 lymphocytes after they have been immune-activated or stimulated to divide. Sarngadharan is now purifying this factor. We know it is a protein much like several other proteins released by stimulated T4 lymphocytes (lymphokines) I have mentioned earlier. We have also learned that a few other already well known factors (such as one called Interleukin-6, or IL-6, and others called TNF, Interleukin-1, and basic fibroblast growth factor) also can help grow these spindle cells; but their effects are not as potent as the new factor. In any case, this T4-cell–derived factor enabled us continuously to grow the KS spindle cell in pure form. Now we could study it in detail.

Our results, some of them in collaboration with Judah Folkman and his colleagues at Harvard Medical School, show that the spindle cell does indeed have many properties of endothelial cells, but also some characteristics of the smooth muscle cells. Since both endothelial and smooth muscle cells form one major structure of the blood vessel wall, it appears, in any case, that the spindle cells behave like an incompletely differentiated cell associated with blood vessels. But that is not what is most interesting now.

Once the spindle cells are stimulated to grow by the growth factor released by the T4-cell, they begin to produce several biologically active molecules, like lymphokines. But because they are not produced by lymphocytes, we refer to them as cytokines (*cyto* means "cell"). These molecules may explain the entire pathogenesis of KS, including additional growth factors acting on the same cells that produce them (although not with as much "force" as the new lymphokine). These additional factors also include some that act to promote growth of normal fibroblasts, endothelial cells, and chemotactic factors (the latter are molecules that draw to the spindle cells other cells, such as lymphocytes and macrophages).

Of even greater interest, some of these factors released by the

*"New" is in quotes because I am not sure the protein has not already been discovered. It could be that we have discovered a new use for a known protein; we will know for sure only when we learn more about its amino acid sequence.

spindle cells promote growth and development of new blood vessels. Some factors were already known to promote blood vessel formation. Owing to research in this field, pioneered years ago by Judah Folkman, at least a few kinds of angiogenic substances (as these growth factors for blood vessels are called) are now known. The blood vessel growth factors released by our cultured spindle cells include some of these known factors.

Our most exciting result came when Nakamura injected some of these cultured human KS-derived spindle cells into mice with suppressed immunity (so that they would not quickly reject the foreign human cells). In these genetically immune-impaired mice, called nude mice, a small mass developed within ten days of injection. Upon examination, it resembled the early stages of KS. When human tumor cells are injected into such animals, the cells often grow as a local mass and sometimes spread; the immune-incompetent mouse cannot reject them, as would a normal mouse.

At first, then, these results did not seem unexpected. We suspected that the KS-like lesion that developed in the nude mouse formed from the growth of the human spindle cells. But to our surprise, the tumor was not human. Examination of the chromosomes of the cells showed they were of mouse origin. Also, we found that when the injected human spindle cells finally died, the mouse's KS-like tumor soon also died out. I interpret these results to mean that the human cells injected into the mouse released factors that produced a mouse KS lesion within ten days!

Another characteristic of KS is that the tumor is *edematous:* that is, much extracellular watery fluid is associated with it. This phenomenon also occurs in the KS-like tumor we can induce in nude mice. Shinsaku Sakurada, a new postdoctoral from Japan working with Nakamura and me, has found that this is caused by an increase in the permeability of the tiniest blood vessels, the capillaries. His results strongly indicate that this is due to the effects of still another cytokine, which he is now trying to identify.

Were all the other biological activities we had found due to already known growth factors and cytokines, or would we find new ones here too, as we had from the stimulated T4-cells we had used to initiate growth of the spindle cells? Studies by Barbara Ensoli in our group show that many of these cytokines are indeed the same as ones already known. For example, the spindle cells manufacture large amounts of the cytokines Interleukin-1 (IL-1), basic fibroblast growth factor, and granulocyte-macrophage colony stimulating factor (GM-CSF). A possible exception, now under study by Thomas Maciag and his group at the American Red Cross Laboratories in Rockville, Maryland, is a growth factor that is related to the known fibroblast growth factors but may differ from them.

In the meantime, interesting independent results came from Gilbert Jay on work initiated while he was at the National Cancer Institute and now being pursued by him in his new position at the American Red Cross Laboratories in Bethesda.* Jay put the *tat* gene of HIV into a developing mouse embryo (this technique of putting a specific foreign gene into the mouse embryo was developed by others; the mice are called transgenic mice†). This procedure allows every cell to have one copy of the gene. Jay found that many of the male mice so treated developed a KS lesion, but these cells did not express the *tat* gene, that is, the *tat* gene was silent in the KS lesion. No *tat* messenger RNA or *tat* protein was made in the KS lesions. Instead, *tat* was expressed by some other nearby cells.

I find these results strikingly parallel to our results. In our case, chronically activated T4-cells (presumably activated in patients by repeated infections and/or by one or more HIV proteins) release a lymphokine that we think acts at a distance to induce a target, presumably blood-vessel–forming immature cells, to become the spindle cells which, in turn, release several cytokines that form the KS lesion as well as the factor(s) that maintain the growth of the spindle cells. In Jay's system, I think that constant expression of *tat* by some cells induces the release of a factor from those cells (or from other cells that the *tat* protein reaches). This factor then acts at a distance to induce the growth of the spindle cells which, in turn, will release the many cytokines we have found.

But we had an exciting surprise ahead, based on a separate hypothesis. I had early on suggested to my co-worker Shuji Nakamura that he should also look into the fluid (media) in which HIV-infected T4-cells

*I have referred to this laboratory before—sometimes as in Bethesda, Maryland and other times, in Rockville, Maryland, as the laboratory moved from Bethesda to Rockville during the course of these events.

†In the transgenic mice every cell of the adult animal has a copy of the inserted foreign gene. Here we refer to the fact that commonly only certain cells will express the foreign gene and make the foreign protein. Which cells express the gene will vary, depending on the particular gene inserted.

were growing. Perhaps he would find that not only activated immune-stimulated T4-cells but also HIV-infected cells might release the same "new" factor. If so, we could suggest two reasons why KS occurs much more frequently after HIV infection: (1) release of the new growth factor by T4-cells because of the chronic stimulation of T4-cells by the ever-present HIV proteins formed from degraded virus and known to be in the circulation, as well as the possible contribution of other microbes that infect AIDS patients with much greater frequency than the rest of the population; and (2) induction by HIV from infected T4-cells of the same spindle cell growth factor(s) we had already found to be released by activated normal T4-cells.

Indeed, we quickly discovered that HIV-infected T4-cells do release some *cellular* growth factor(s) for spindle cells, but the amount released is very low, and I cannot yet state that this is the same as the one released by activated but uninfected T4-cells. What was much more exciting came from the findings that much of the growth factor activity released by the HIV-infected cells came out only early after new T4-cells were infected. My co-workers Ensoli and Wong-Staal knew that the HIV *viral* protein, *tat,* was made only after acute infection of cells. We were also familiar with the results of Alan Frenkel, then at Johns Hopkins and now at the Whitehead Institute in Boston. He had made the amazing finding that the *tat* protein could bind to and enter cells. In light of this result and our earlier results and those of Jay, it remained for us to show that *tat* could also be *released* by HIV-infected cells and that it too could be a growth factor for the spindle cells.

This is precisely what Ensoli quickly found. She also found that the concentrations of *tat* required to promote spindle cell growth were very low, a remarkable result for many reasons. To my knowledge it is the first time a protein of a virus, normally working as a regulator for virus formation, was observed also being released from the infected cell and acting as a growth factor for another cell. In some ways this is reminiscent of the *tax* protein of HTLV-1 and HTLV-2, which is a regulator of the virus and yet also influences growth of cells. But in the case of *tax,* it was not, as far as we knew, by acting directly as a growth factor (and as far as I knew, not by being released by the infected T4-cell), but by turning on the gene for a cellular growth factor (IL-2) and its receptor within the HTLV-infected T-cell, as I have already discussed.

As a marvelous example of how science moves, however, brand-new results from John Brady and his co-workers at the National Cancer Institute and by a new postdoc in my laboratory, Ron Gartenhaus, have found that *tax* can also act as an external growth factor. This concept—the release of a viral protein, particularly a regulatory protein (one made by a viral gene generally for the purpose of helping the virus replicate in the cell but not incorporated into the virus) and its capacity to influence the behavior of nearby *uninfected* cells—is new to virology. I propose it will prove to be a major determinant in the way many viruses produce their disease-causing effects.

What about the greater incidence of KS in HIV-infected homosexuals compared with other HIV-infected groups and the male dominance in all forms of KS? As I have already implied, these findings appear to require another environmental (external) factor—such as another not yet discovered virus—in gay men, and this may yet turn out to be the case. Since the male dominance also occurred in Jay's mice experiments, however, these results do not require another environmental factor for their explanation. Perhaps this is a result of differences in the levels of male sex hormone, testosterone. The epidemiological studies that have shown KS to be more frequent in gay men than in other HIV-infected people have mainly compared the infected gay men to IV drug addicts and hemophiliacs. Perhaps sexually active homosexuals as a group have higher testosterone levels than do male intravenous drug addicts and hemophiliacs.

This is a testable hypothesis. It is also attractive at the cellular-molecular level because sex hormones can affect the production of cytokines and their interaction with cell receptors. As such they may significantly influence growth and behavior of the spindle cells. This too is testable, and it is testable hypotheses that advance science. Some recent results are consistent with this idea. For instance, in collaborative studies with us, Prakash Gill at the University of Southern California found that testosterone promotes the production of IL-6 from the spindle cells, and IL-6 can augment spindle cell growth.

Perhaps also gay men have a greater incidence of T-cell activation than do IV drug addicts or hemophiliacs. If so, they will secrete more of the new lymphokine, which is such a potent growth factor for spindle cells. This will soon be tested in our laboratory. Of possible significance in this regard, Mark Kaplan of North Shore Hospital on

Long Island (an affiliate of Cornell medical school) reported at my last annual lab meeting two cases of KS in IV drug addicts. Both were infected not only with HIV-1 but also with HTLV-2. As I have said, T-cells infected by HTLV can be a source of the same lymphokine. If HTLV-2 is playing a role here (as opposed to just a coincidental infection), then it must be by a mechanism that is augmented greatly by the presence of HIV (the additional growth factors induced by HIV are needed) because infections by HTLVs alone do not lead to KS.

But we have a way to go before we fully understand epidemic KS. For example, a recent epidemiological result is troubling. Harold Jaffe of the Centers for Disease Control analyzed the occurrence of KS in HIV-infected women who had had sex with HIV-infected bisexual men and compared the result with the incidence of KS in HIV-infected women who had sexual contact with HIV-infected blood transfusion recipients or with HIV-infected intravenous drug addicts. Presumably they acquired HIV from these sexual partners. KS developed far more often in the women who had sex with the HIV-infected bisexual men than in those who acquired their HIV infection from HIV-positive drug addicts. Although the study was limited in its numbers, should such results be verified the implication is obvious: another factor—presumably a microbe, and presumably one that is sexually transmitted—is involved in the origin of epidemic and perhaps other forms of KS, which thus far we have not found.

To summarize my thoughts on the development of KS, I think it originates from a less than fully differentiated cell that has the capacity to form blood vessels when it is stimulated by certain cytokines and lymphokines, and when it is located in the appropriate architectural environment. The cell is called the spindle cell when found in a KS lesion. Several cytokines induce growth of the spindle cell. The most potent one is released by T4-cells when such cells are immune-stimulated, as may occur in some forms of chronic infection. This lymphokine is currently being purified and produced in large amounts in our laboratory for more detailed studies. Once the spindle cells are actively growing they make and secrete many well-known cytokines, and it is these cytokines that produce many of the characteristic features of the tumor we call KS. In other words, a cascade of events is triggered after the first step—the promotion of growth of the spindle cell. In HIV-associated KS there is an additional chronic stimulant of spindle cell growth. Each time new T4-cells are infected by HIV, the viral protein called *tat* is transiently released by the infected cell. At very low concentrations *tat* enhances spindle cell growth.

There are still other ways HIV infection may enhance spindle cell growth. As we have seen in chapter 13, it is known that there is an abnormal activation of a large number of B lymphocytes after HIV infection, even though these cells are not infected. Some workers have indicated that this may be an effect of the HIV envelope protein gp120, released as HIV particles are degraded. It is known that activated B-cells release the cytokines Interleukin-6 and *Tumor Necrosis Factor,* TNF,* and as I mentioned, both of these molecules also have some growth-stimulatory effects on spindle cells. To add more oil to the fire, TNF may enhance HIV replication, giving rise to more HIV-infected cells—and so the cycle goes on.

KS can sometimes be a reversible tumor in its early stages because in this phase it is not a real cancer. The cells are not autonomous but rather depend on the supply of stimulating cytokines (lymphokines) and/or *tat* for keeping the spindle cells growing. In this period the KS lesion resembles the normal growth healing that occurs after a skin abrasion. After chronic abnormal cell growth, however, there is a high probability of one or more genetic mutations occurring in these spindle cells which, in some cases, lead to their autonomous growth. At that stage the KS lesion develops into a true malignancy called a sarcoma. At this stage the tumor is no longer reversible by depriving it of its triggering agents.

The results derived from the experimental systems that have led to this discussion were originally carried out only in the HIV-associated epidemic KS. We thought that similar mechanisms might be involved in the development of other forms of KS, and recently we had the opportunity to study some of the endemic (African) KS as well as

*This protein is so named because when it was first discovered its effects included the induction of changes in and around some animal tumors, which led to the death of some of the tumor cells. We now know, however, that the effects of TNF are far more diverse. TNF is released from a variety of cells, and it is important to us here because it is present in abnormally high amounts in patients infected with HIV; it can promote more HIV formation; more HIV infections promote more TNF production, as Tony Fauci and his group have shown; and it is also produced by the KS spindle cells, and it adds to the growth and promotion of spindle cells (as we have found) and produces effects on other cells that contribute to the development of this complex tumor.

some cases of classical KS and KS developing after kidney transplantation and immune-suppressive therapy. Our results suggest that, again, the key is spindle cell growth and release of cytokines—in other words, the same mechanisms may be at work in all types of KS. The key to the origin of KS in my mind is anything that can contribute to the increase in number and activity of spindle cells. Drs. S. Sakurada and S. Nakamura in my laboratory find that corticosteroids (like cortisone) greatly enhance the ability of the T4-cell lymphokine (and some of the other, lesser effective growth factors for these cells) to promote spindle cell growth.

Earlier I noted that transplant patients receiving immune-suppressive therapy sometimes develop KS. Corticosteroids are part of that therapy, and when such therapy is terminated the KS lesions sometime permanently vanish. I contend that the development of KS in these people is due not to immune suppression but rather to the immune-suppressive therapy's triggering of the immune system (T4-cells) to make and release the lymphokine that promotes spindle cell growth. In the setting of steroid therapy, this growth promotion is marked. This hypothesis too is testable. We already know that steroids do augment the lymphokine effect. Now we need to develop assays for serum levels of these spindle cell growth factors to see whether they are increased.

I have no useful ideas about the rare development of KS in older Mediterranean men or of endemic African KS. I assume, like most students of these diseases, that the initiation of the former involves a genetic susceptibility and the latter an infectious agent (yet to be discovered). I would speculate, however, that spindle cell growth factors will be involved in both.

The evolving KS story is a good example of how our laboratory works on a medical problem. I think it is also a good example of how our group can be most productive. Only after the development of useful experimental systems can real progress be made in understanding and often in treating human disease. It has been said that ideas are cheap—it is work and getting answers that count. While this is hyperbole to make a point, the point is not without merit. In the absence of useful systems to study KS, we had been left only with epidemiological and clinical studies or anatomical examinations, and many speculations. These can go only so far. Now we have tissue culture laboratory systems for the study of what seems to be the key cell of KS, and we have the mouse animal model to explore other questions.

We are now beginning to use these systems to test compounds for their efficacy in reducing growth of the spindle cells in the tissue culture system. When a compound appears of interest in this *in vitro* system, we move to the mouse model and test whether the compound under study can delay or prevent development of the KS lesion or, after the tumor is induced in the mouse, whether the compound can effectively destroy the tumor without serious side effects to the animal. We can use these systems in an empirical manner in an attempt to find effective drugs for KS—a trial-and-error approach. But based on some of our findings with these new systems, we can also rationally search for or design drugs likely to be effective. We can try, for instance, to develop inhibitors of the lymphokine or of *tat,* which promote spindle cell growth. We can also develop compounds that interfere with the *effects* of the cytokines released by the spindle cells.

Some commentators on the history of science have noted that scientists did not need to understand any fundamentals or mechanisms to come up with antibiotic cures of some bacterial diseases, nor did Sabin, Salk, and Koprowski need to know how the polio virus worked to develop their vaccines. This is true. But I think that for the most part those days are gone forever. Most accidentally or readily solvable problems have *been* solved. Vaccines that depend simply on finding an organism have mostly been developed. Increasingly, benefits will likely come from basic understanding of disease processes, which in turn come from the basic advances in disciplines like molecular biology, immunology, and cell biology. And as I have illustrated, these in turn must be used in laboratory systems that offer the possibility of asking—and answering—the right questions.

15

About Causes of Disease (and, in Particular, Why HIV Is the Cause of AIDS)

When are we ready to say that we know the cause of a disease? More precisely, when do we have sufficient information to say this or that is a cause of a particular disease; a contributing factor (co-factor) essential, along with the true cause, for the disease to occur; enhancing the progression of the primary cause; or irrelevant to the disease? To a greater extent than we might want to believe, there are few hard-and-fast rules to help us make these determinations, certainly no cookbook recipe to follow. I believe what we do use is very much the sauce of common sense heavily spiced with the right experiences.

In some cases the cause of a disease or sickness is so evident that debating it would be absurd. Traumatic injuries are examples. In other cases, the origin of a disease is so complex or multifactorial that, rather than talking in terms of a cause, it seems more reasonable, as in *some* human cancers, to think of increasing numbers of contributing factors making something likely to occur with higher and higher probability: without the presence of any one of them, the disease would occur with less frequency, though the graphed line is not a straight one. A third category (the most common) lies somewhere in between. There is a cause, that is to say there is a *sine qua non,* and often there are contributing co-factors that increase the probability of the disease developing, its severity, its speed of onset, or all of these.

Alternatively, other factors may contribute to a lessening of the probability of a disease developing. Most infectious diseases fit into this third category.

Let us take the extreme case of trauma. We would not get much of an argument about the cause of a serious head injury to a bicyclist who has collided with a moving truck. We do not need statistics; common sense rules. But suppose there were four cyclists in the collision, three wearing red shirts and two wearing protective helmets. One of the latter lands on wood and is unharmed; the other on concrete and is hurt. The two not wearing helmets also land on wood. Both are hurt. The three hurt were all wearing the red shirts. Again, we do not need a statistical study. We would hypothesize that the helmet was likely protective, at least when one lands on wood; and that without a helmet, one receives more serious injuries on concrete than on wood. Despite the perfect correlation, we would not think too much about the red shirts.

We would not be able to prove our hypotheses from these four bits of data, but with enough time and numbers, they would be borne out. Because of our previous experiences, it makes sense to us: concrete is harder than wood; injuries are worse when the part of the body receiving an impact is not protected; an unprotected skull is more vulnerable than a helmeted skull. We could argue that the cause of the head injury was concrete, the red shirts, the absence of a helmet, or the truck—but we don't. The impact of the truck is the *sine qua non,* the cause. The others are influential positive and negative factors or correlations with no influence at all, as in the case of the red shirts. If, however, the correlation with red shirts continued in the future with dozens of additional examples, obviously we would have to suspect that, even without an immediately logical reason for color to influence such injuries, it would have to be considered and studied as a possibility.

Only in the area of infectious diseases does medicine have formulated guidelines for determining causality, and famous ones at that. They are called Koch's postulates, or Henle-Koch postulates. Robert Koch and Jacob Henle formulated these notions about one hundred years ago, not because they were pontifical or dogmatic persons but in order to stop the tendency of some physicians to lay claim to having found a cause of every disease with every new finding of a microbe. Rules were needed then, and can be helpful now, but not if they are too blindly followed. Robert Koch, a great microbiologist, has suffered from a malady that affects many other great men: he has been taken too literally and too seriously for too long. We forget at times that we have made great progress in the last century in developing tools, reagents, and diagnostic techniques far beyond Koch's wildest fantasies. What remains constant is the need for common sense and clinical and lab experience in working with disease in general and with the particular disease in question. I believe Koch would have been the first to recognize this.

Koch's postulates, while continuing to be an excellent teaching device, are far from absolute in the real world outside the classroom (and probably should not be in the classroom anymore except in a historical and balanced manner). They were not always fulfilled even in his time. Certainly, they did not anticipate the new approaches available to us, especially in molecular biology, immunology, and epidemiology, or the special problems created by viruses. They were, after all, conceived only for bacterial disease, and even here they often fail. Sometimes they are impossible to fulfill; many times one would not even want to try to do so; and sometimes they are quite simply erroneous standards.

Briefly stated, these are Koch's famous postulates:

1. The germ must be found in the *affected* tissue (the presumed site of the disease) in *all* cases of the disease. (At every site of disease tissue, we find the germ.)
2. The germ must not occur in other diseases or as a fortuitous nonpathogenic agent. (In every case where we find the germ, we find the disease.)
3. It must be repeatedly isolated in pure form, free from other germs and from the host body. (We must be able to identify and study the germ on its own.)
4. The germ so isolated must be able to induce the disease anew and be re-isolated from the inoculated animal. (Every animal into which we introduce the germ must get the disease and yield new germs.)

It is important to emphasize again that these were *ideal* guidelines for demonstrating a *bacterial* cause of disease *one hundred years ago.* Even then Koch himself knew some of their limitations and did not always demand their fulfillment in his own work. (Somehow these points are lost on the few who argue against HIV as the cause of AIDS.) For example, Koch became very famous after finding the bacterium of tuberculosis. Though he isolated the microbe in pure form and could reproduce the disease in an animal, his hypothesis was flawed. First, TB, we now know, appears in many different clinical forms. (I suppose this is because of dose, route of transmission, and genetic factors of the person or of the microbe.) Sometimes TB is a disease of the spine, other times a kidney disease, but most of the time it is in the lung. More important, most infected persons are *healthy carriers.* Koch's postulates did not consider this notion, which is a central theme today in the study of almost all infectious diseases. In short, even for his classical work on TB, postulate number two was not fulfilled.

Further, we do not always isolate the bacterium from every case of TB. Some scientists say this is because other diseases resemble TB, that the negative findings are not really TB cases. But this is not true. Any experienced medical intern can vouch for the many false negatives for the microbe before a successful isolation. Experienced pathologists can testify to the failure to isolate the bacterium until the postmortem, when the diagnosis is often easier, especially when the pathologist accidentally becomes the recipient of the germ and inadvertently fulfills one of Koch's postulates by coming down with the disease.

Koch's postulates also had problems accounting for cholera. Being a brilliant scientist, with self-confidence and an obvious competitive spirit, Koch was in a bit of a race with Pasteur to find the cause of this new or re-emerging deadly epidemic. He discovered the cholera bacterium. He "knew" it was the cause of cholera. It was repeatedly associated (correlated) with the epidemic outbreaks, but he could not produce cholera in animals. Max von Petenkofer of Munich argued with him and, to prove Koch wrong, swallowed about one-third of a teaspoon of the microbe. He got diarrhea—but he did not die of "cholera." He must have died thinking he had forever disproved Koch, but today we know that only a fraction of people infected with the cholera bacterium actually get cholera. In fact, subclinically infected—and totally asymptomatic—carriers are the rule. And this is not an aberration restricted to cholera. *Expression of disease is the exception for the majority of microbes.*

There was still another problem with the postulates. Friedrich Löffler, another pioneering German microbiologist and Koch's associate, discovered "Löffler's bacillus," the causative agent of diphtheria. But he found the microbe in a membranous lesion of the throat of a sick child. Unfortunately, the disease was not of the throat but of the brain, and there was no microbe there! By the standards of Koch's postulate number one, Löffler had not found the cause—but later work showed that he actually had: that the microbe in the throat releases a *neurotoxin,* a substance that gets into the bloodstream and harms the brain, an *indirect* but critical effect of the microbe.

The case of leprosy may be even more illustrative. Clinical leprosy is not hard to diagnose, but even today the bacterium is elusive, and as far as I know we are still not able to produce leprosy in animals. Koch's postulate number four is not fulfilled. One hundred years after Koch, we know one hundred times this many exceptions. In some cases we know clearly why only a fraction of infected people develop a disease when exposed to the microbe identified as its cause. In most we do not.

Viruses make for even more trouble. A virus *cannot be isolated and grown outside cells.* It can be grown only in cells. Bacteria and other metabolizing living microbes can, of course, usually be grown in pure form in one rich broth or another, and these microbes can often produce disease in many different animals. *Viruses' trait of growing only in cells frequently restricts their host range.* It is often impossible to find an animal that a given human virus can infect. An inability to produce a bacterial disease in animals had even occurred in Koch's work in a disease for which he knew the cause—the case of the cholera bacterium—but this was probably an exception with an accepted bacterial cause of a human disease. With viruses, a restricted range of target sites is the norm. *Many viruses do not produce disease even when introduced into an animal; in many instances this is because they cannot under any circumstances infect the animal.* Does EBV, the cause of infectious mononucleosis, produce this disease in animals? Does human hepatitis B virus cause hepatitis or liver cancer in animals? The answers are no. How many times has influenza been produced in animals with any of the many different influenza viral strains?

The case of AIDS is also illustrative. The receptor for HIV is, as we have seen, a very specific molecule on the surface of some human cells, and receptor binding is only the first requirement for successful infection. HIV cannot even be tested for disease production except in chimps and gibbon apes. They are the only animals capable of being infected in any reproducible manner. (We severely limit such tests, however: gibbon apes are endangered, and chimps are nearly so.)

I often remind my co-workers that even when we can succeed in infecting animals with a human virus, we can get a misleading result. I have already noted in an earlier chapter the powerful example of the gibbon ape leukemia virus, which can produce leukemia when given to young gibbon apes but produces no disease when inoculated into other primates. Conversely, *herpes saimiri* does not produce disease in its "normal" host monkey, but when injected into some other kinds of monkey it can induce a cancer of the lymph glands (lymphoma). The conclusion is that no conclusion can be drawn from such studies in the absence of other information to help us in our attempt to decide on a causative role for a virus.

The last one hundred years have also conclusively taught us that the underlying premise of Koch's postulates is wrong: it assumes that there is one causative agent for each disease, and that any one disease has only one causative agent. We know better. For instance, one of the hepatitis viruses also sometimes causes cancer years after infection. It may or may not, just as it may or may not bring on an acute hepatitis soon after infection. I have already mentioned the many different manifestations of the TB bacterium. Everyone knows syphilis produces an acute genital ulcer, and the microbe can then be found and isolated. Some remember that later it can produce skin abnormalities, and years—sometimes many years—later it can cause crippling heart and/or brain disease, and we may not be able to find the microbe then. (The mechanisms involved for all three are not exactly the same.) Measles virus causes common measles; long after infection, it can cause a brain disease. Papilloma viruses can cause common skin warts, but some strains (with not-yet-identified co-factors) can cause genital and cervical cancer after many years.*

We have seen that HTLV-1 can cause leukemia after many years, but it can also cause neurological disease and very likely several other

*Papilloma viruses causing skin warts are actually a very interesting example of a case where common sense tells us we have the cause of a pathology—we see them in the wart with regularity, and animal experiments have shown their ability to produce these lesions. But the virus has yet to be isolated, even by virus standards (that is, successfully transmitted and grown in cells in the laboratory). Hepatitis B virus and the leprosy bacterium have also yet to be grown in laboratory culture.

entities. Polio virus infection very early in life usually causes nothing—but if infection is delayed and the patient has not received a vaccine, a crippling attack on the nervous system ensues. Smoking causes sinusitis, bronchitis, cancer of the larynx, cancer of the lung, and emphysema. It increases the incidence of cardiovascular disease, cancer of the breast, cancer of the bladder, and death by accidental fire (only this last one occurs quickly). Smoking, then, is another obvious example of one agent acting as a cause and/or contributing co-factor to many diseases.

Our lessons are even clearer in animal virology. Here we can control which animal we inject and when. We have seen in an earlier chapter, for example, the myriad diseases induced by the feline leukemia retrovirus. Most of the time we do not know why a particular disease results; sometimes the onset of disease correlates with dose, route of entry, strain of virus, and with the genetics of the host or some other co-factors.

Sometimes we have a virus in search of a disease. This was the case after my co-workers and I discovered human herpes virus-6 in 1986. It was ubiquitous in most populations. But did it cause any human disease? Subsequent epidemiological and immunological studies in Japan have provided strong evidence that it causes a relatively benign childhood disease. Other evidence suggests that if infection is delayed to teen or adult life, it may produce a disease similar to an infectious mononucleosis-like disease. We can safely conclude, then, that *one agent can cause more than one disease.*

In the century since Koch, we have also seen the opposite: *one apparent clinical disorder can have more than one cause.* For example, there is no doubt that leukemias of animals and humans can have more than one cause, even leukemias of the same cell type. In fact, this may be true for every human cancer.

Perhaps the most important developments since Koch's time have been in our technological approaches to a "new" microbial disease. We have longitudinal and cross-sectional epidemiology, and we have sophisticated tests for antibodies in the sera of infected people—including the ability to subclassify the type of antibody. Some antibodies, called *immunoglobulin M,* or IgM, occur shortly after infection. Others, called IgG, occur later. This allows us to follow a healthy person and, if his or her serum is taken at the right time, to estimate the time of infection and correlate this with disease signs and symptoms.

More recently, molecular epidemiology has given us a more complex but even finer tool. It allows us to probe with considerable sensitivity various tissues and body fluids for the presence of the viral genes, that is, for active and/or inactive virus. Finally, we now have a number of laboratory cell culture tools to study the effects of a virus, and to determine from these studies whether the target cells of the virus and its effects in these cells make sense with the clinical picture we think the virus may produce.

Other virologists have remarked on the failings of Koch's postulates, especially for viruses. More than fifty years ago, T. M. Rivers proposed the need to add "the development of a specific immune response to the agent." Al Evans, the noted Yale epidemiologist, recalls that Rivers also emphasized the notion of the *healthy carrier,* dismantling a key support to Koch's postulate, and Robert Huebner added the element of the longitudinal and cross-sectional epidemiological studies. Evans himself has suggested we apply these immunological criteria: absence of antibody in sera before disease and before exposure; antibody regularly appears during illness (both IgM and IgG); antibody before exposure indicates immunity (I don't agree with this one because it doesn't always hold true); and antibody to no other agent should be similarly associated with the disease unless that agent is a co-factor.

HIV is the sole primary cause of the epidemic called AIDS. This was our conclusion in early 1984, based on:

1. The finding of a new virus in AIDS patients either by serum antibody testing (in 88 to 100 percent of cases—varying in different protocols in the initial studies) or by isolation of the virus (in about 50 percent of cases). Isolates were made from blood cells, plasma, bone marrow, and thymus glands.
2. The virus was also found in a very high percentage of then so-called "pre-AIDS cases" and in the groups known to be at high risk for developing AIDS, but only rarely in healthy heterosexuals in the United States.
3. The characteristics of the virus showed it was a new, or at least not a known, virus. AIDS as an epidemic was clearly new.
4. Wherever the HIV was found, AIDS was present or soon present. Conversely, no HIV—no AIDS.
5. Studies of blood donors and their recipients (done in collaboration

with CDC investigators) showed that nineteen of nineteen blood recipients who developed AIDS (with no other known risk factors) had HIV, and at least one of their blood donors was HIV-positive (at a time when fewer than one in two thousand donors in the United States were infected with HIV). The study was soon slightly expanded to twenty-eight of twenty-eight, another perfect correlation.

6. The virus infected T4 lymphocytes. We soon discovered that HIV also infected macrophages, the cells central to immune function.
7. We commonly found HIV in the brains of people who had died of AIDS. Dementia is another common feature of AIDS.

Soon CDC and others, including my lab, reported evidence against other agents (microbes or otherwise) as serious candidates for the cause of AIDS. Thus, it was by a process of both inclusion and exclusion that HIV came to be known as the cause of AIDS.

What have we learned since? Results would accumulate, verifying and extending the earlier results and making the data for HIV as the cause of AIDS one of the most compelling we have for the cause of any disease by a virus. I will mention just a few examples.

In country after country, as AIDS was recognized in an epidemic form, it was always noted *after* HIV was present. In other words, the development of AIDS in every country studied has correlated with the entry and subsequent increase in HIV in that population.* And the estimated incidence of AIDS development after HIV infection has steadily increased. In 1990, it appeared that more than 50 percent of all infected people would develop AIDS, probably much more. Despite its slow course, HIV is one of the most effective viral pathogens known.

Over the last few years large prospective† epidemiological studies of homosexual men showed a major statistical correlation in development of AIDS in those who become HIV-positive (or were so) versus those who remain negative.

HIV has now been isolated in close to 100 percent of cases with

*As a new epidemic virus in all geographic areas, it is not hard to show this. Sera stored from many geographic areas that were studied decades earlier will show none or only a very rare positive for antibodies to HIV.

†A *prospective* study is one done before certain events occur—in this case, following HIV-negative homosexuals with no disease for years, *then* determining the presence of HIV and disease.

AIDS (a far greater efficiency than we have with most other viruses already accepted as causes of disease). By tests for virus isolation, for antibody in sera, or for the genes of the virus (by molecular probing), virtually *all* cases of AIDS can now be called HIV-positive, whereas the healthy, "nonrisk" groups have very low rates of detection.

A longitudinal prospective study of twenty-two children born to HIV positive mothers from Bordeaux, France, by D. Douard and colleagues revealed that all had antibodies to HIV when tested shortly after birth. Since mothers transfer their own antibodies to their infants *in utero,* this is to be expected. Such antibodies last several months, then decline and disappear—unless the infant has been infected with virus *in utero,* during delivery, or by nursing. If infected, the infant will in time develop his or her own antibodies. By one year, fifteen of the twenty-two babies showed a decline, and then complete loss, of antibodies. Virus was not isolated from any one. All remained healthy. Antibodies remained in the sera of the other seven, however. *Antigenemia* (viral proteins in the blood reflecting presence of virus) were found in all, and HIV was isolated from four of four tested (one died and two infants and their mothers did not return to the clinic). All seven of these children developed signs and symptoms of AIDS. In short, the uninfected infants of HIV-positive mothers in no case developed AIDS. The HIV-infected ones did.

We also have much more information on pathogenesis, that is, on how HIV exerts its disease-causing effects. Increasingly, this is a virus that *makes sense* in explaining the basic abnormality in AIDS that sets the stage for so many other infections and problems. Since I have already discussed some of these points, I will deal here only in response to specific arguments some have brought to this issue, arguments which I believe to be false.

The vast majority of people who have raised, re-raised, and re-re-raised objections to the conclusion of an HIV cause of AIDS seem to have little or even no experience in science or medicine: for example, Jad Adams, an English writer, who wrote *AIDS: The HIV Myth* (1989); Katie Leishman, who writes for *The Atlantic* (a magazine that is, I think accurately, described as both moderate and mainstream) and has been a proponent of syphilis as the cause of AIDS; and Anthony Liverridge, who writes repeatedly in favor of almost anything he thinks is against HIV. One other AIDS "writer" is an aficionado of an African pig virus. He believes it is all a CIA plot. A frequently quoted New York physician named Joseph Sonnabend has repeatedly promoted multiple causes for this one new epidemic, favoring the idea of multiple risk factors. Some scientists—I guess myself more than others—are targeted by these writers, perhaps understandably. What is not understandable is their lack of concern or insight. With the exception of Sonnabend, they have not prepared themselves in any way that would enable them to pronounce on human diseases. The situation would be comical if it had not to some extent affected human life.

Many serious scientists, *prior* to the May 1984 publications, had proposed candidate causes of AIDS, including antigen overload—*no specific cause,* but too many infections harming the immune system; a fungus (ruled out, as I mentioned earlier, by the lack of any data and by the logic that AIDS has been transmitted along with Factor VIII, a substance used to treat people with hemophilia, which in its preparation is filtered in a way that would eliminate microbes the size of a fungus); a mycoplasma*; and many other agents. But since May of 1984 there has been wide acceptance in the medical and scientific community of HIV as the indisputable cause of AIDS.

This has been acknowledged, for example, in the United States by the Centers for Disease Control, the National Institutes of Health, the National Academy of Science, and the Institute of Medicine. It has also been the conclusion of the World Health Organization and, to my knowledge, of every health and biomedical scientific agency, foundation, or society in the world that has expressed any opinion on the cause of AIDS.

**Mycoplasma* are like viruses, rickettsiae, and chlamydia in that they are small enough to pass through an ultra-filter (450 nanometers). They differ from them in being able to have their own complete metabolism and reproduction outside of cells. In this sense they are more like the larger microbes, such as bacteria, fungi, and parasites. Unlike bacteria, they do not have a rigid outer cell wall but only a membranous envelope. This is one reason for their particular characteristic of extreme variability of form—both size and shape.They were first isolated in 1898 by Edmond Nocard and Pierre Roux in association with a fatal pneumonia of cattle known in Europe since the seventeenth century. In humans, one type causes a pneumonia. More often, mycoplasma are a most unwanted but common accidental laboratory contaminant of cell culture systems. Recently—and to me astonishingly—they have been proposed as a primary co-factor not as an opportunistic infection, but as a real co-factor in AIDS by Montagnier, who now further suggests that they are the chief cause of the death of the T4-cells, based mostly on his recent results detecting some in his culture of the AIDS virus. Montagnier is also recommending treating AIDS with antibiotics—which do inhibit mycoplasma and which he finds also inhibit his T4-cell culture death when used with blood containing HIV and also apparently carrying mycoplasma. I do not agree with the recommendation, on the basis of our work or on what evidence I have seen; nor do I think it has logical merit. Nonetheless, this surprising view, which has been chiefly presented in press conferences, has given, and may do so for a while, added longevity to confused and confusing (to others) arguments that HIV is not the primary cause of AIDS.

One scientist, however, Peter Duesberg, has become almost famous for promoting in the media and in a few scientific journals his very strong opinion that HIV is simply a tagalong passenger. AIDS is caused by a lifestyle of risk behavior. . . . There is no cause . . . there are a lot of causes. . . . If a person is HIV-infected, it doesn't matter. . . . If he infects someone else, it doesn't matter. . . . Certainly, one shouldn't take antiviral AZT medicine, because AZT is immunosuppressive and would therefore help cause AIDS. Duesberg's notions have influenced a few scientists, such as Robert Root-Bernstein, who also has written about the subject, and Duesberg's longtime colleague at Berkeley, Harry Rubin, a senior laboratory virologist.

Root-Bernstein, a physiologist at Michigan State University, has rewritten many of Duesberg's arguments and is in agreement with all of them. He is particularly adamant about the need to fulfill Koch's postulates yet seems to be unaware of their limitation or of the fact that neither Koch nor his co-workers always fulfilled them. Like Duesberg, Root-Bernstein is unaware that not fulfilling those postulates is the rule and *not* the exception for diseases caused by viruses in general.

He offers still more points against HIV and for the lifestyle theory. Normal cells ("normal" people), he says, resist HIV infection, as demonstrated by the fact that HIV is "only grown in a peculiar cell line." HIV-infected AIDS patients have immune suppression for *other* reasons and, when manifested, then and only then can they be infected by HIV. This is manifestly untrue. No virus would survive if it infected only an immune-suppressed person. In any case, HIV easily infects (in the lab) "fresh" healthy macrophages, T4-cells, and Langerhans cells obtained directly from normal healthy donors. Further, of course there are HIV-infected but not yet immune-suppressed people—prior to the onset of AIDS. This is documented every week if not every day by some lab or another. We have even seen examples where this happened to a technician working in an HIV production facility who was accidentally exposed to and infected by HIV.

Root-Bernstein also thinks that as HIV-infected people progress to development of the disease, they have less and less virus, and that this argues against an HIV causation. Both the fact and the logic are false. Newer and well-known data show HIV levels rise with disease progression, and arguments based on how much virus there should be at any particular time (early or late) are pure sophism. He states that 5 percent of people with AIDS are antibody-negative and that this is an argument against HIV. It is not. Lack of antibody development may occur for many reasons. The key is whether there is *virus.* With improved methods of isolation and identification, HIV can now be found in all patients with AIDS.

He then argues that there is just as good a correlation of AIDS with herpes simplex virus (HSV), Epstein-Barr virus (EBV), cytomegalovirus (CMV), and hepatitis B virus as there is with HIV. This is not so. The correlation is *almost* as good. But the key point is that these (especially HSV, EBV, and CMV) are old and ubiquitous viruses. Most of us are infected with them. It is almost the same logic as saying that AIDS correlates just as well with the presence of hair. HIV is relatively new, at least for most populations, and is obviously not ubiquitous. Its correlation with AIDS obviously means far more, and it is far more specific.

Root-Bernstein goes on to say (in an as yet unpublished thesis) that the "one thing we can be certain of . . . in the failure of most healthy animals not being susceptible to HIV infection is that there *first* must be immune suppression for infection to occur." This is surely a big leap for a man arguing for caution in our ideas. Obviously, the logic is flawed. There is no reason to assume that the failure to infect is caused by the fact that the animal is not immune-suppressed. There are reasons that should be glaringly apparent to a microbiologist, such as the animals' not having the appropriate receptor on the surface of their cells. Why does he think chimps and gibbon apes get infected? Is it because these animals are always immune-suppressed? No. It is at least in part because they have a CD4 molecule on the surface of some of their cells that is similar to human CD4.

Like Duesberg, Root-Bernstein also believes blood transfusions themselves cause immune suppression and that this accounts for the association of AIDS with blood transfusion. In this point he ignores or doesn't know that thousands of blood transfusions performed daily all over the world are not producing AIDS. He may also be unaware of the fact that sometimes just a few units of blood have given AIDS; and, to my knowledge, when blood transfusions have produced AIDS, at least one of the donors was found to be HIV-positive. The correlation of blood-transfusion–induced AIDS with HIV contamination in the blood is as close to perfect as anything I have seen in medicine.

After making some points on Kaposi's sarcoma with which I agree (we do not understand all the factors that contribute to KS development—HIV alone may or may not be the sole agent in epidemic KS, and if HIV is indeed a major factor, as I believe it to be, its role must be indirect), Root-Bernstein tells us that (1) there is evidence for specific chromosome abnormalities in KS, and (2) they are not consistent with the role of any virus. He is wrong on both counts. Chromosome changes in KS are varied—often no abnormalities are seen, and the kind of chromosomal abnormality seen in a given cancer does not tell anything about whether the change in cells of the tumor is due to a virus, a chemical, some other physical cause, or sheer chance.

As for Duesberg, he has a Ph.D. and is a Professor of Molecular Biology at the University of California in Berkeley, with a background in chemistry. He made very significant contributions to our understanding of the molecular biology of animal (especially chicken) retroviruses many years ago and is a member of the National Academy of Sciences. On the other hand, he is not an epidemiologist, a physician, or a public health official. More important, to my knowledge Duesberg has never worked on any naturally occurring disease of animals or on any disease of humans, including AIDS. Nor, I believe, has he ever worked with HIV.

Duesberg's arguments have been modified over time, as should anyone's as new information becomes available. He has so far published the following as his major points:

1. AIDS is not new, not contagious, and not a single disease. The U.S. government (CDC) hastily adopted HIV into the definition of AIDS. Thus, with that definition one must have HIV to have AIDS.
2. AIDS is not really a syndrome or a new disease but a collection of many other, older diseases and factors that result in immunodeficiency. These include malnutrition, blood transfusions, lifestyle, and drug abuse (including antibiotics). Babies with AIDS get the drugs *in utero* from their mothers. AIDS in babies is different from that of adults anyway, so it must really be a different disease.
3. Immunodeficiency cannot account for tumor development such as that with Kaposi's sarcoma and lymphomas.
4. AIDS correlates only with "risk factors" and not with HIV. When HIV is found to correlate, it is simply that: correlation, not causation.

5. HIV violates classical conditions of viral pathology. Viruses cause disease quickly and after much virus replication. In contrast, HIV is correlated with disease only late after infection "when the virus is not replicating" and when antibody is present, and antibodies protect against virus.
6. These diseases associated with a decline in number of T4-cells (AIDS) could not lose these cells due to a retrovirus because retroviruses are never cytopathic. Some can cause cancer. Others do nothing. HIV is a simple, unassuming retrovirus possessing no AIDS-specific or special genes.

   Despite claims to the contrary made by Daniel Zagury and myself and by David Ho, Mark Feinberg, and David Baltimore, Duesberg states that HIV does not kill T4-cells. He uses a report by Schnittman to support his statement that 99 percent of the few T4-cells ever to be infected survive as reservoirs. Not enough T4-cells are infected to be killed off by HIV. In fact, HIV is really a dud.
7. The long time between infection by HIV and disease precludes the virus as a cause of AIDS.
8. Koch's postulates are not fulfilled:

   a.) Many people who have HIV infection still do not have AIDS;

   b.) HIV is not present in the tissue site of disease (the critical example he uses is that HIV is not in brain neurons, yet brain disorders occur);

   c.) HIV does not produce AIDS in animals; and

   d.) Some people with AIDS did not have HIV (for example, he quotes the work of my lab, reporting in May 1984 roughly a 50 percent success rate in isolating HIV from AIDS cases).
9. Babies get AIDS because they are exposed to the same things their mothers are exposed to (drugs, for example). It doesn't matter that AIDS in babies correlates with HIV.
10. Similarly, it doesn't matter that hemophiliacs and other blood recipients get AIDS correlating with transmission of HIV, because blood itself is immunosuppressive.
11. Duesberg discounts the SIV monkey model as a disease too removed from human AIDS. He says the disease occurs too rapidly to be compared with AIDS and in other ways is unlike human AIDS.
12. He believes a wide chemical spectrum of drugs causes this new epidemic: from cocaine, heroin, and poppers to antibiotics, anes-

thetics, and AZT, to multiple overexposures to microbes and even to simple blood transfusions (without any microbe).

13. He also states that blood-transfusion AIDS in the United States doubled between 1988 and 1989. If HIV is the cause, testing for it (which began in 1985) and not using positive blood should have brought about a decline.

It is hard to argue with him and others about AIDS being old and being many, many different diseases. How can I refute this? One can only point out that Duesberg is a chemist, a molecular virologist. He has never studied a human disease, now or in the past. No physician, no epidemiologist, no health worker from any part of the world to my knowledge would agree with this view. Of course, immune suppression has occurred in the past, and especially the recent past in association with some drugs, radiation (surprising that he left this factor out), and pronounced malnutrition. The argument that old, common infections produce this kind of disease is simply untrue, as is the argument that AIDS occurred by simple blood transfusion. It is hard for me to believe that anyone visiting several New York hospitals today could argue against AIDS being a new epidemic.

Duesberg has trouble with the concentration of AIDS in young gay men (in the United States and Europe), but HIV does concentrate in this group. He says viruses don't select by sex, but he is wrong: some do, and for many obvious and logical reasons. In fact, HIV is not the only virus more prevalent in this group. So are hepatitis B virus and some other sexually transmitted microbes. If enough members of a group are in more intimate physical contact than others then, of course, there will be more infections in the group, particularly if the virus has much greater ease in its transmission by way of blood (more blood penetration is likely to take place in rectal sex than in vaginal sex). We know HTLV-1 females are more likely to have been infected by sex with men than are HTLV-1 men to have been infected by women; in fact, the least risk of sexual transmission is to a man by a woman. The same is true of HIV. Changing the probabilities of transmission are co-factors I mentioned earlier, such as venereal ulcerating infections. These lesions in a woman apparently allow greater ease of HIV transmission to a man, and I think, as Frank Plummer and others have reported, this is a major factor in female transmission to males in parts of Africa.

It seems to me preposterous to argue that antibiotics and anesthetics play an important role in this epidemic. There is no epidemiological data to support any such assertions. The notion of babies getting AIDS from their mother's drugs is almost as preposterous. When a woman has two babies and one gets infected with HIV and the other does not—what happened? Such examples do exist, as well as the telling study from Douard and colleagues in Bordeaux that I mentioned on page 285.

With regard to the tumors that develop (KS and lymphomas), I do not much disagree with Duesberg. I do not think simple immunosuppression is the answer, and I have given my views and a summary of our experimental results on KS in the previous chapter. But the current evidence strongly argues that HIV infection increases the possibility of KS development, and we think we are beginning to learn why.

Earlier Duesberg had argued that HIV was not correlated with AIDS (as Root-Bernstein continues to do), but apparently his objection is not as vigorous now. He now accepts the correlation, but concludes that correlation does not equal causation. True enough, but it is one hell of a good beginning. It is especially powerful when both virus and epidemic are new in most populations and when *prospective* epidemiological studies show clear linkage, as they have for HIV.

Although I have dealt with Koch's postulates (as an argument against HIV and more generally) earlier in this chapter, I would like to discuss them again here in relation to Duesberg's arguments against HIV. They include the following:

1. Many people who have HIV do not have AIDS. He is right—but most, and possibly all, will. Apparently forgotten is the asymptomatic carrier and subclinical infections known for almost every microbe. We learned this in the century since Koch, and actually Koch knew it himself. Think of how many people get exposed to and infected from Koch's TB bacterium and never get sick!
2. Conversely, he states that tests that find antibodies to HIV indicate earlier exposure and current protection but not continuous infection (the suggestion is that the presence of antibodies indicates the virus is being killed). This is a startling argument because Duesberg has experience in retrovirology. He must know that most antibodies are *not* protective. In many instances none are protective. Protective antibodies, or those we call neutralizing antibodies,

are often absent, or too low, or the virus escapes them by mutations—which is just what occurs with HIV! In any case, HIV isolation since 1983–84 has improved. Today most labs can isolate HIV from almost any case of AIDS, and certainly detect it by serum antibodies or molecular probing in all or virtually all cases. I know of no better linkage in any viral disease.

3. The next Koch postulate Duesberg uses is the requirement that the microbe be in the *diseased tissue.* Since dementia is often a part of AIDS, since brain cells called neurons are the key to thinking, and since HIV does not infect these cells, this is a failing of the evidence that HIV causes AIDS. I have already discussed that AIDS is not the first disease to fail to meet the demands of this postulate, and used the illustration of diphtheria, a case in Koch's own lab. The microbe was found in membranous material of the throat; the disease was in the brain. Later the medical world learned why.

   The microbe produces a neurotoxin that gets into the blood and when it reaches the brain produces its damage. Most ironically (and rather amazingly) Duesberg ignores the fact that my co-workers and I demonstrated HIV in the brain six years ago—a finding that has been extensively confirmed. Yes, it is true that later we showed it was chiefly in microglial cells of the brain, and probably some of the other cells making up the very architecture of the brain and providing links between one part and another. Obviously, the damage to the cells could have adverse effects on neurons. There is also reason to believe that infected microglial cells may inappropriately release cytokines, molecules that might have deleterious effects on neurons.

   Finally, as I have noted briefly in chapter 13, Volker Erfle and a co-worker in Munich have recently shown that one of the HIV regulatory proteins, *Nef,* can be released from infected brain cells. They report that *Nef* has toxic effects on other (uninfected) brain cells and, interestingly, shares some structural features in common with the neurotoxin from scorpions. The other hallmark of AIDS is, of course, immune impairment, and who could think of more appropriate target cells for a virus to infect and impair in the immune system than T4 lymphocytes and macrophages?

Discounting the monkey model and the monkey virus, SIV, was unfair. Duesberg must know his statement that the disease is much more rapid is due to the fact that the scientists *selected* for variant SIV strains and most susceptible animals in order to obtain results in large enough numbers without having to wait five to fifteen years. But in fact, there are now examples of SIV taking years for disease to be detected, just as with HIV in humans. His discussion of the blood transfusion results is even less fair. The increase in pediatric AIDS in blood-transfused infants or children is due to the time it takes for AIDS to develop: that is, for the most part they became infected *before* the blood test was available. Since these are *children who are HIV-positive,* the argument is nonsensical.

As I have already noted, Duesberg's harshest criticism of HIV as the cause of AIDS has been his contention that it "violates classical conditions of viral pathology" (whatever those are). He believes all viruses cause disease quickly and with much virus replication, whereas with HIV infection, AIDS occurs late, when there is no HIV replication and when antibody (presumably protective) is present. He thinks that no retrovirus is cytopathic, that HIV does not kill T4-cells, and that not enough HIV is found in the body to account for the disease.

All of these points are wrong. Many viruses cause disease after a prolonged period. That they might not be like "classical" virus infections means only that medical science has learned new things in the past few decades. Slow virus diseases are now well known. Encephalitis of measles, for example, occurs years after the infection; at the time of onset, infectious virus is only rarely found. Kuru and Creutzfeldt-Jakob disease often occur decades after exposure to the strange agent that causes them, as does their counterpart in animals, the scrapie agent of sheep.

Examples of slow disease by viruses abound in animal retrovirology, a remarkable slight on Duesberg's part, since he has been a molecular retrovirologist. Feline, avian, and bovine leukemia viruses may produce leukemia years after infection. Similarly, HTLV-1 in humans usually takes a decade or longer to express itself. Other classes of human tumor viruses take just as long: hepatitis B virus and liver cancer, certain strains of papilloma virus and genital cancer, among others. Another example comes from viruses known as *papovaviruses.* Some of these small DNA viruses can induce cancer in animals. A human counterpart, the JC virus, can produce a disorder known as progressive multifocal leukoencephalopathy, or PML (the name we all prefer). This is a serious and chronic disease of the brain which, too, can occur long after infection and only under special conditions.

But this discussion of slow viruses would be incomplete if I were not to mention that among the most typical (perhaps we can even use the word *classical*) slow viruses are certain lentiretroviruses of the hoofed animals. These viruses have been known for decades. Distantly related to HIV, they produce disease in animals after a latency that can last several years. Duesberg simply dismisses them by saying "they probably don't cause the disease." This is an amazing statement: virus can and has been deliberately inoculated into host animals and has reproduced the disease!

These "critics" do not realize that some viruses infect, establish themselves, and can persist *despite* the immune response to them. This is, of course, patently obvious in the case of retroviruses whose DNA forms stay in the chromosomal DNA. Or, if they do realize this point, they think the virus cannot reestablish its replication once the immune response is generated. But the evidence is clear that the immune response *fails* to control HIV. Variants emerge that get around the defense. Neutralizing antibody and cellular immunity simply do not suffice.

In respect to the claim that there is little or no virus as disease progresses, all the data I am aware of over the past five years say just the opposite. Work by Eva Marie Fenyö and Birgitta Asjö in Stockholm, Jay Levy and co-workers in San Francisco, Leibowitch in France, Ho in New York, Schnittman at NIH, and many more clearly indicate that the amount of HIV increases as disease advances. Some of the results go further: HIV amounts increase *before* disease advances, and more actively replicating variants begin to dominate.

Another contention of the anti-HIV people is that no retrovirus is cytopathic (cell-damaging or killing), a statement they uphold without ever having studied HIV! In fact, Temin showed years ago that some avian retroviruses (a subfamily of the very class of viruses Duesberg worked with) were cytopathic, and anyone who has worked with HIV knows it surely is. Now it is true that some isolates are not nearly as cytopathic as others. I know of no study that shows that there are not cytopathic variants present in any AIDS patients. Our experience in LTCB is that every isolate we have studied has in one way or another harmed cells of the immune system.

I have also emphasized the indirect effects of HIV. Whereas I agree with these critics that autoimmunity has not been shown to play a role in AIDS development (that is, the immune response to the virus mistakenly attacks human cells as well, particularly cells of the immune system), there are other indirect viral effects that in my mind are suggestive of an important role. Remember that HIV infection can lead to an inappropriate release of cytokines, powerful molecules that, when secreted in the wrong place and at the wrong time, may produce many of the features of the disease. I have also discussed the effects of viral proteins from degrading virus, such as the envelope of HIV that inhibits T-cell growth and function. Also recent evidence from MIT scientists indicates that the *Tat* protein of HIV inhibits the normal response of a T-cell to activation by an antigen.

In addition, we can state (as Howard Temin, William Blattner, and I have already done in a letter in 1989 to *Science* magazine) that, to show cause of disease by a microbe, one need not have to know and show how the microbe produces its disease. Almost all past attribution of causality, including those of Robert Koch, were without such explanations. The final irony is that we probably know more about how HIV produces its pathology than about the pathological mechanism of virtually any other microbe, and what we have learned fits with the disease manifestations.

Like any student of HIV, I am well aware that numerous other microbes and factors may contribute to HIV disease. Obviously, the opportunistic infections produce their own specific problems—for example, the pneumonia produced by the opportunistic microbe pneumocystis. I concede that the Kaposi's sarcoma story, too, is complex. I agree that certain fungi and many viruses produce a large, variable, not always predictable clinical course. Many of the clinical disparities (and there are not nearly so many as some people argue) between different HIV-infected populations depend on the local microbial flora or different ages of infection.*

But these are the side effects and the consequences of HIV infection. It is HIV that is the *sine qua non* for the epidemic. I also intuitively agree with the idea that co-factors *for HIV progression itself* can also exist. In an earlier chapter and, in fact, for the past five years, I have suggested we look carefully at two: HTLV-1 and HHV-6. There may be more. Without them, however, AIDS will still occur. It will

*The course of disease in some newborns is much more rapid than in other HIV-infected groups. It appears to depend on the age of infection. If the newborn was infected *in utero* early in pregnancy, when there is little development, the disease is more rapid.

probably just take longer. Nonetheless, I am surprised that Montagnier has suggested a mycoplasma (also known as pleuropneumonia-like organism, or PPLO) as a possible or probable *prerequisite for AIDS development with HIV,* based on his few detections in culture samples. In short, he has lent *some* support to Duesberg (who, interestingly enough, dismissed Montagnier's idea). Montagnier thinks this common laboratory tissue culture contaminant is necessary for the T4-cell depletion in AIDS. I think that this is unlikely.

The biomedical scientific community has varied in its response to Duesberg. Most scientists have kept mute and get some fun out of what he says. Some who have expressed a view see Duesberg as a useful gadfly, to paraphrase Harvard's Jerry Groopman, even though they see him as very wrong. Those within the media who have given him a forum have been openly laudatory. Others say that even if he is wrong, at least he has made us realize how little we know about HIV and how it works—but I think we know plenty about HIV and how it works, though of course we need to know more. Other commentators have been far more critical of Duesberg's stance, including Robin Weiss, former director of the Chester Beatty Cancer Institute in London and longtime retrovirologist, and George Klein, professor of immunology at Stockholm's Karolinska Institute. I tend to agree with the opinion expressed openly by David Baltimore that Duesberg's rush to the media has its dangerous side. His voice has diminished confidence in science and scientists, medicine, physicians, and health care workers.

Undermining confidence in the only people working on AIDS is not likely to help unify our efforts to conquer it. Instead, it breeds distrust and perhaps frank hostility, often turning researchers into convenient targets for those whose business it is to attack government institutions and people. As I look back on medical history over the past fifty years, I see few examples in which the medical-scientific establishment arrived at an important conclusion about a disease and was later shown to be wrong. Irresponsible rantings against HIV as a cause of AIDS may lead some infected people to forgo anti-HIV therapy or, worse, not to care about whom they might infect.

When I am asked to explain why these critics do not accept the scientific results we have achieved with HIV, I reply that there actually are people (I hope not many scientists) who do not believe the United States placed a man on the moon. There is also, I am told, a Flat Earth Society, which has evolved a complex rationale to explain away all the evidence that the earth is round.

16

What We Can Do About AIDS and the AIDS Virus

The title of this chapter touches, of course, on the most pressing and important issue in this book. I address the issue by discussing the responses first of individuals and society and then of the medical scientific community, breaking the latter into therapy for the HIV-infected individual and prevention of HIV infection.

Much has already been written about the social, economic, legal, and ethical aspects of AIDS by people with more front-line experience than I will ever have and, therefore, with far greater authority. I will only briefly add my agreement with the more common sentiments. In that we can be certain that AIDS will not be completely solved in the next few years, we must have long-term programs on sex education. We cannot be bored by the repetition of the task because there will continue to be new people at risk. Sex education is a central aspect, but I do not believe we are 100 percent sure of what constitutes safe sex. We are told that there is a substantial decline in the rate of HIV infection among some homosexual groups. But even if this is more than just a saturation phenomenon, we cannot conclude that it is chiefly the result of the use of condoms. It could be much more due to a behavioral change—a decrease in the number of sexual encounters.

Just as important as promoting the use of condoms will be fostering continued education on the nature of the virus and an improvement in science education in general. I don't believe there is a shortcut to reach these difficult goals. The greatest need for education is among those many feel are uneducable: drug addicts. We have little choice but to try, however, and some groups, such as those at the University of California at Davis, believe they have indications that it can be done. The issue of whether to adopt a uniform policy of giving out free sterile needles and syringes makes sense on the one hand, and there are experiences that indicate that it works. For example, there was a reduction of infection in Glasgow, where they were distributed, in contrast to Edinburgh, where they were not. On the other hand, I cannot help but feel that such a policy is tantamount to giving up on drug addicts, and there are those who believe it encourages the use of drugs in some communities. What we can all do is to become far less tolerant of a drug culture, one that, for example, laughs at jokes about cocaine (perhaps more so in the early period of the cocaine epidemic, but even now) on national television, in movies, and in homes that foster the acceptance of drug use.

All of us can also vocally support aid to less privileged nations, particularly those with a major AIDS problem. From even a totally selfish viewpoint, it would be worthwhile to be doing more for these people. For disease spread by microbes, there are no longer any isolated populations. For viruses of some types, what can occur "there" may eventually occur "here." There has been significant organizational response to the needs of central Africa in battling AIDS. The World Health Organization (WHO) is the largest in terms of the number of people, the intensity of work, and the funding. On the other hand, I am more than a little worried about the future of WHO, since it lost the head of its AIDS program, Jonathan Mann, who in my opinion has been one of the most effective leaders I have come across in the fight against AIDS. Smaller organized units are or will soon be similarly involved, such as the World AIDS Foundation (funded by the royalties from the blood test patent by mutual agreement of the United States and France) and the World Laboratory Project in virology (funded mostly by Italy).

Sweden's new program also merits mention. It is funded by SAREC, its agency for research in underdeveloped nations, and is targeted toward specific AIDS-related projects that will provide information and experience of immediate use for a few African nations. The objectives and plans of SAREC are exemplary for their long-term commitment, medical scientific focus, and care in interactions with colleagues in the African countries.

But parts of Africa, parts of Brazil, Thailand, Haiti, New York City, and some other regions will need more help than this. There are areas of high HIV prevalence where blood tests are not even performed and where medical education related to AIDS is sorely lacking. Some of these regions need such medical knowledge and intervention more than anywhere else. It comes as a shock for a laboratory scientist to learn that standard vaccines for diseases such as measles are not administered in many regions of Africa because of insurmountable logistical and economic problems. So the prevailing feeling is that the same would be true if a vaccine for AIDS were developed. But the differences between these diseases are enormous. For this to happen with AIDS would be an unforgivable failure on the part of the world public health effort and of governments in both developed and underdeveloped countries. The very thought of failure to disseminate such a vaccine properly would ensure the greatest concerted public health mobilization efforts targeted to a region of the world in medical history.

Individuals can even play a role in helping the cases of AIDS in underdeveloped nations through long-term expression and concern for the problem. It is the individual who, by raising the consciousness and the conscience of society, can shape attitudes that will stimulate greater involvement.

The contributions of medical science toward controlling AIDS and the AIDS virus can include diagnostic, therapeutic, and preventive procedures. Although the diagnostic blood test for HIV is now readily performed in most places, the development of a simpler assay with more rapid results and a much cheaper cost would benefit underdeveloped nations greatly. However, I am certain this kind of work is not receiving adequate attention—probably because there is little financial incentive for it.

People with AIDS need both primary and secondary treatment. By *secondary* treatment I mean supportive care, such as replacement of fluid and minerals lost with diarrhea, replacement of blood if there has been hemorrhage, and attacks on complicating infections and tumors with antibiotics and antitumor agents, respectively. By *primary* treatment I mean direct attacks against the virus by antiviral chemotherapy or biotherapy (nonchemical, such as interferon and other cytokines) and/or significantly overcoming the immune abnormalities. I am uncertain about the prospects for so-called chemical immune stimulant agents used to produce a general stimulation of the immune system. In the absence of control of the virus, these agents may only spread it. Also, I do not know what some people mean by "immune modulators," even though they are already being clinically used in parts of the world. Supposedly such things make the immune system function better. The term is, of course, vague, but it sounds good. Precisely what "modulation" occurs? I know of no chemical that has been shown to have a general, beneficial immune-restoring effect, and I would be suspicious of any claims to the contrary.

Clearly, what one aspires to in helping the immune system is the restoration of T4-cell number and function, correction of any macrophage dysfunction, and promotion of both cytotoxic T-cell killing of cells expressing virus and B-cell production of antibodies that neutralize virus infectivity. In other words, we want specific interference with virus replication and specific promotion of immune attack—the former with inhibitors of virus release and of virus infectivity, the latter with a vaccine.

In theory there is an approach to the restoration of immune function that might work: the complete eradication of the bone marrow of an infected person and its replacement with bone marrow from a normal individual, while simultaneously treating the recipient HIV-infected person with intensive antiviral therapy during the early period of infection. One form or another of this approach has been suggested by a few clinical immunologists, such as David Sachs at the NIH and Shimon Slavin of the Hadassah Medical School in Jerusalem, but serious attempts have so far been held back by the fear of the obvious dangers inherent in the procedure in a person already immune-compromised by AIDS. It would be more likely to succeed in the early, preclinical phases of HIV infection, but in this case the physician faces the perennial dilemma: Is it worth the risk of harm to an individual who is not yet experiencing any medical consequence of HIV infection?

Before leaving the area of "immune modulation," I should recall the odd story of cyclosporine. Like the unexpected arrival in the night of a long-lost relative, announcements of a clinical breakthrough in AIDS therapy by the use of cyclosporine, a compound that *inhibits* immune function and particularly T-cell function, suddenly appeared. The three French clinical immunologists who made the announcements were presumably influenced by the hypothesis put forward by Montagnier that much of the pathogenesis of AIDS was likely to be the result of an HIV-induced autoimmune disease.* If this were the key to AIDS, then it might make some sense to further destroy the immune system, albeit with great risk because of the immune suppression already existing in an AIDS patient. The problem, of course, was not limited just to the risk but included the fact that there was and still is little evidence that any major abnormality in AIDS is the result of autoimmunity. The oddity with cyclosporine, however, was more in the manner of its presentation than in the idea or the science behind it: as rapidly as the claims of major benefits were made (the patients were but a few and the time but a week or so), they disappeared.

In contrast to the limitations in approaches affecting the immune system, there are many approaches for interfering with the virus. The essential purpose of any primary treatment of an HIV-infected person is to stop the spread of virus. To be as effective as possible, treatment should begin as soon after infection as possible and continue for life. Since most antiviral drugs for AIDS patients will have side effects, once again there is the dilemma. If not all HIV-infected people develop AIDS (and that is a possibility), is it justifiable and wise to treat someone who shows no symptoms in early HIV infection on the basis of the notions that serious disease is probable and that early antiviral treatment is likely to be much more effective than later treatment? I would say yes, depending, of course, on the evaluation of the toxicity versus the anti-HIV potency of the drug in question. Each drug has to be individually evaluated in this way, as does, sometimes, the patient.

Once the primary role of HIV in the cause of AIDS was established, some concepts of inhibitors could be developed based solely on our understanding of the replication cycle of animal retroviruses and some of the modifications introduced with the HTLVs. It was imperative, however, to create an experimental system for actual testing of HIV

*An *autoimmune disease* is an attack on one's tissues by one's own immune system. For example, in AIDS the notion is that part of the disease is the result of an overlap of the immune system's attack on HIV with components of normal cells. Although we can find some antibodies to HIV that do interact with one or another cell (as is common in many other diseases), there is no evidence whatsoever that this interaction results in cell damage or impaired function.

inhibition by various compounds. The development of a reproducible tissue culture system for continuous replication of HIV (the H9 cell-line clone and other cell lines infected with various strains of HIV) paved the way for this as well as for the elucidation of some of the significant and unique features of HIV replication destined to yield additional ideas for antiviral treatment.

When the CD4 molecule, present on the surface of the T4 lymphocytes and macrophages and certain other cells related to macrophages, was identified as the principal cell receptor for HIV, the idea of using free CD4 molecules to block the binding of the virus to the cell was conceived by several groups. The concept called "molecular decoy therapy" utilizes purified CD4 to compete with CD4 on cell surfaces for binding to the virus. The complex of CD4 and virus would eventually be degraded. Mass production of pure free CD4 by recombinant DNA technology was achieved a few years ago by at least three commercial groups. As anticipated, laboratory *in vitro* tests showed that it inhibited virus entry into cells. Clinical therapy was initiated in 1988 by several groups, particularly Samuel Broder and his colleagues at the National Cancer Institute, Paul Volberding in San Francisco, Jerry Groopman in Boston, and a team at Duke University. The results to date have revealed little or no toxicity, but despite the promising results reported by Norman Litvan in the SIV-infected macaque, there are, disappointingly, no indications of significant effects in reducing HIV or in clinical improvement in HIV-infected people. This is almost certainly due to the great difficulties in achieving and maintaining adequate blood levels of CD4.

I think that, because of the possibility of a major success by the CD4 therapy approach, many more clinical and laboratory studies designed to maintain and increase blood levels of the effective moiety of CD4 should be carried out.* It is also of interest to consider the problem of drug resistance here. This occurs commonly with bacteria and with cells (cancer cells) treated with agents designed to kill them or to block their growth. As we will see, resistance can also be a problem with antiviral therapy. But it should not occur so easily with CD4 therapy. HIV resistance to CD4 therapy should mean that HIV has mutated to a virus that can no longer cause AIDS, for the interaction between CD4 and virus is in my mind the most essential aspect of the HIV-AIDS story.

*Unfortunately, later studies still show puzzlingly disappointing results.

The steps involved in virus penetration after binding to CD4 are poorly understood. I know of no effective inhibition at this level. There are, however, other compounds reported to interfere with virus entry whose mechanisms of action are poorly understood and for which there is no logic for any specificity in their effects. One of these is a carbohydrate polymer known as dextran sulfate. Another goes by the commercial code name of AL 721. I do not think that either will have an important future in therapy of HIV infection because of the nonspecific interactions of both and their minimal *in vitro* inhibitory effects.

An obvious target is the enzyme reverse transcriptase. Its unique aspects compared with cellular enzymes, its essential function in virus replication, and the ease with which inhibitors of it can be studied make it almost ideal. Also, there is considerable experience to draw upon with RT from other retroviruses. It is not surprising, then, that the earliest investigations designed to find effective drugs against HIV were with RT inhibitors. The first was with a compound called suramin. The studies were carried out initially by Broder in collaboration with some investigators in my group. Although the laboratory results were impressive and showed some margin of safety (they blocked virus at concentrations that did not show toxic effects on cells), the clinical studies proved that the drug was too toxic to be safely used.

Work at the Pasteur Institute soon led to clinical trials with another compound with a coded company name, HPA-23. It became very famous when Rock Hudson went to Paris to receive it. His trip stimulated several other Americans to head to Paris for treatment. I think it is still in clinical use today but without publicity. It seems doubtful to me that Rock Hudson or anyone else benefited very much from HPA-23. It is a polymer, specifically a *polyanion,* a large molecule composed of negatively charged repeating units. It is a heteropolymer because the units repeated are not identical; they consist mostly of negatively charged molecules of tungstate. Years ago some workers at the National Cancer Institute in Bethesda working with Chermann found that this substance, like many other chemicals studied earlier and since, had some inhibitory effects on RT of the mouse leukemia retrovirus in tissue culture tests. Unfortunately, the HPA-23 effects are marginal on the RT of HIV, and it has little or no inhibitory effects on HIV infection of cells in laboratory culture.

AZT is another story. It is a powerful inhibitor of the reverse transcriptase step in HIV replication. The studies by Samuel Broder and his co-workers at the National Cancer Institute (originally in collaboration with Dani Bolognesi and his colleagues at Duke) were systematic prior to its clinical use, and the claims for *some* clinical effectiveness were made only after careful pharmacological studies were completed and after clinical tests showed significant differences in survival and morbidity (incidence of disease) between placebo-treated and AZT-treated AIDS patients.

The mechanism of the AZT effect on RT is known. As the viral RNA is transcribed into DNA the DNA chain grows, one purine or pyrimidine nucleotide building block after another. In an earlier chapter I made the analogy to a pearl necklace, each pearl being a nucleotide. Picture the necklace opened and growing pearl by pearl. Each pearl has a hook at one pole to attach it to the growing necklace; another at its opposite pole receives the next pearl. AZT is a pearl without the second hook. It attaches to the growing DNA chain, but it cannot accept the next nucleotide. Thus it is called a *chain terminator.* Of course, our cells make new DNA before they divide, and they need enzymes (DNA polymerases) to catalyze this reaction which, although different from viral RT, is related to it in several respects. AZT, therefore, will also have a side effect, like most drugs.

But there is some margin of specificity. Probably because the virus replicates more than do our cells, and very likely also because RT is more error-prone (it uses wrong nucleotides more frequently than do our enzymes that catalyze synthesis of cellular DNA), considerable inhibition of HIV infection without cell toxicity occurs at lower AZT concentrations. As one would anticipate, the side effects at higher concentrations are chiefly of cells that, like those of the bone marrow, are dividing and, therefore, making DNA. So AZT might cause a decline in the number of certain blood cells. For this and other reasons, AZT therapy must be carefully followed by people with experience with such agents.

A second problem is the development of resistant variants of HIV. This too is expected. A third problem is its high expense. Nonetheless, AZT is the one agent that has been put through careful clinical trials and shown to work. It is not the answer but it does prolong life and, in the main, reduce morbidity. It also has reversed neurological damage in some cases. I think the work of Broder and colleagues partly in collaboration with David Barry at Burroughs-Welcome with

AZT is already in the category of classic studies of rational therapy of a human disease, but in the end this work probably will be most remembered for the true commencement of a new phase in medicine: antiviral treatment, or perhaps "antiviotics."

AZT stimulated studies of related compounds which, like AZT, are analogous to the DNA purine and pyrimidine building blocks. Most of them, like AZT, probably work by chain termination. Some that show promise are dideoxyinosine (ddI) and dideoxycytosine (ddC). There are other ways of inhibiting the RT reaction, which I will not go into here except to mention that some such compounds are available and under clinical study. Phosphonoformate, for instance, inhibits HIV RT. It also inhibits the DNA-synthesizing enzyme of some herpes viruses, such as HHV-6.

Another HIV-specific enzyme is a protease which HIV makes and then uses to "cut" its core proteins into smaller and functional sizes. This is a required step in the HIV cycle of infection. Our cells also have proteases, but much is now known about the chemistry of the HIV protease, and differences from the cellular proteases have been found. This opening has led many groups, especially those in the pharmaceutical industry, to predict and to make compounds that inhibit the viral enzyme. This seems to be one of the most promising areas of HIV research.

Other parts of the HIV replication cycle invite our attack, such as the complex steps employed in the expression of the integrated DNA provirus back to viral RNA, the subsequent transport and splicing of this RNA, the maturation of viral proteins from larger to small structures, and the assembly and release of the virus at the cell membrane. As already indicated, the RNA step requires the function of virus-specific proteins (*Tat* and *Rev*),* and the protein cleavage requires a virus-specific enzyme (the viral protease). No inhibitors of *Tat, Rev,* or the protease have yet shown clinical efficacy, but this is one of the most active and important areas of current research.

Some of us are also considering new, untested, but logically attractive approaches to antiviral therapy based on molecular biology. One of the more actively pursued approaches of this kind is called *antisense RNA.* When a gene is expressed it is, of course, in the form of a messenger RNA molecule (mRNA) that is "translated" in the cell

*The genes are written *tat* and *rev;* the proteins, *Tat* and *Rev.*

cytoplasm by the capacity of the mRNA to interact with cell components, in particular with another class of RNA known as transfer RNA (tRNA). Each tRNA can carry a specific amino acid, the building blocks of cell proteins. As I have briefly summarized in an earlier chapter, the mRNA provides the ordering of the amino acid by calling for different tRNA molecules as its "script" is being "translated." It does so in order of three nucleotides (the *codons* of the genetic code), that is, a triplet of nucleotides calls for a certain tRNA, then the next triplet, and so on, thus determining the amino acid order. The amino acids form bonds that link one to another. Their order will determine the nature of the protein. The same process is true for viral RNA and proteins.

If the nucleotide sequences of key viral mRNA molecules are known, and we do know them for HIV, then we can synthesize a portion of the complement of those sequences in the laboratory.* This nucleic acid fragment is not translated. It will not make a protein, but it will bind to the active messenger RNA. In so binding, the viral mRNA will not be translated to make viral protein. In laboratory tests such approaches have been shown to work in a collaborative study involving my colleague Prem Sarin with Paul Zamecnik and John Goodchild of the Worcester Foundation for Experimental Biology in Massachusetts. Related but independent studies by another colleague, Flossie Wong-Staal, and still others by Sam Broder and his co-worker Hiroaki Mitsuya, have shown the same efficacy: HIV replication is blocked.

But there are significant problems to overcome for the clinical application of antisense RNA, mostly in the technology of sufficient production but also in achieving the right blood levels and getting enough of the nucleotide sequence across the cell membrane barrier. If this can be achieved, in theory it should be a highly specific and useful approach. It also illustrates some of the unique problems for AIDS investigators and the media. Wong-Staal had presented her findings at a spring 1989 scientific meeting in China. Some of the Chinese and Italian press were present. Oblivious to her report, I was speaking elsewhere at the time. I was suddenly confronted with excitement and questions about our antisense RNA treatment which, unbe-

**Complement* here refers to the fact that certain nucleotides stick with other nucleotides. This pairing occurs by a process called hydrogen bonding. For example, UACG will pair with an AUGC sequence because U forms hydrogen bonds with A, and G forms them with C.

knownst to me, was already headlined in parts of Europe as a major advance in AIDS therapy. The reality is that it *might* be, someday. For the moment it is an interesting laboratory result that will be pursued.

For some years now we have frequently discussed but never tried a far more unorthodox approach. We and others have been able to make defective mutants of HIV. Some can infect and replicate but have no visible cell pathogenic effects. Others infect and express some of their proteins but cannot replicate. Some in my group have gently argued for taking bone marrow cells from an AIDS patient and deliberately infecting them with these defectives. The idea is that such cells will not be harmed but may instead be permanently protected against being infected from the HIV in other cells of the individual. Theoretically, such cells might go on to proliferate and restore immune function for the person. The worry is whether such defectives will indeed prove to have been harmless. How can we know? Also, could they recombine* with the HIV in the infected person and lose their defectiveness?

Another related unconventional approach is the possible use of gene therapy for HIV-infected people. This is not the place to discuss the technology or feasibility of gene therapy, but I will mention some of the principles as they apply to the problem of HIV infection. Gene therapy has not yet been successfully used for any human genetic disease, but as the precise gene abnormality becomes defined for more and more genetic disorders and the efficiency of replacing the abnormal gene with a correct one continues to improve, it seems quite likely that this could become a routine in the not-too-distant future.† Oddly enough, it will probably be by the use of an innocuous animal retrovirus that will carry the "good" gene to be inserted. In this case the retrovirus is referred to as a vector.

In addition to the problems of efficiency of transfer of the "good" gene to enough of the right cell types for some disease, there is also the problem of delivery to the right tissue. AIDS, of course, is not a genetic disease. Here our goal would be to fool, destroy, or block the

**Recombination* is a well-known phenomenon of retroviruses in which portions of genetic information of two different things can combine. The "head" of one chromosomal strand, for example, may combine with the "tail" end of another and vice versa.

†During this writing the first case was tried. The disease was due to a specific mutation in a gene leading to a defective enzyme, adenosine deaminase. Early results by French Anderson and Michael Blaise of NIH indicate that replacing that gene with a normal gene may already be providing some chemical improvement for the patient.

function of some HIV gene or gene product (protein) necessary for HIV to replicate. The possibilities are numerous. For instance, Eli Gilboa at Sloan-Kettering Institute in New York is experimenting with transferring the genetic sequences that *Tat* binds to. His approach would be to take bone marrow of an HIV-infected person and transfer the nucleic acid sequences (called *Tar*) to which *Tat* binds.

As we have seen, when the HIV DNA provirus of an infected cell is turned on, *Tat* and *Rev* are the first HIV proteins made. These "early" proteins are essential for the formation of the other HIV proteins. *Tat*'s role is in its interaction with special regulatory sequences (*Tar*) of HIV that positively influence the formation of functional viral messenger RNA. *Tat* will bind to the incoming *Tar* sequences instead of to the "normal" *Tar* present in HIV because the former will be made in excess. Others have long considered related approaches, such as transferring mutants of *Tat* that can bind to these *Tar* sequences of HIV, but the mutant *Tat* cannot carry out other steps of *Tat* function. Since these functions are essential for HIV to complete its cycle of infection, there will be no further HIV replication. When a mutant protein dominates the normal viral protein, we refer to it as a *trans* dominant mutant. These and many more like them are exciting ideas today. I imagine they will be exciting antiviral therapies tomorrow.

The final approach I think worthy of mention is antiviral immunotherapy, the use of a vaccine to stimulate an anti-HIV immune response, not to prevent infection but to hold back the virus in an already infected person. This approach was first initiated in humans by Daniel Zagury and his co-workers in Paris in 1986. Using specific viral proteins presented in different ways to the patient at roughly three-month intervals, his results, with Odile Picard at the San Antoine Hospital in Paris, are the most detailed and impressive to date. In 1987 Jonas Salk, working with Alexandra Levine of the University of Southern California, began related work, employing killed whole virus (usually missing a component of the viral envelope). Like the Zagury-Picard studies, these approaches are designed to enhance cellular immunity. They too show promise.

To develop a real vaccine for AIDS means to make something that stimulates an immune response that will prevent infection by HIV, or of disease development if infection cannot be blocked. The approach does not fundamentally differ from vaccines for widely different microbes, but some of the difficulties with HIV are unique. Most of the successful vaccines against viruses have been with a *live attenuated whole* virus or a *killed whole* virus.* I already mentioned most scientists' aversion to this for HIV, and FDA's almost certain aversion to its use in an uninfected, healthy person. So the first handicap for an HIV vaccine is the probable need for a *subunit* vaccine, one or more viral proteins, or a fragment of such protein(s).

But this need not be insurmountable. The Merck Company, for example, made a successful subunit vaccine for hepatitis B virus. There is more than one way to make viral proteins or pieces in sufficient amounts for a vaccine. One is by chemical synthesis (synthetic vaccine). The other is by one of many different approaches from recombinant DNA technology. For example, putting the gene for the desired viral protein—let us say, the gene for the envelope of HIV—in some form that keeps it active so that the protein can be made in large amounts. The gene must also be in a form† that can be transferred into and replicate in a bacterium or some alternative carrier. If designed properly the protein will be secreted from the bacterium into the extracellular fluid from which the investigator can purify the viral protein and then use the purified protein as a candidate vaccine.

In other approaches the entire microbe is used in delivery of the viral protein. This can be done with some harmless bacteria. In addition to the delivery by a bacterium of the gene for the candidate viral protein vaccine, some investigators have used another virus already in use for vaccine purposes. For instance, we began a collaboration with Bernard Moss of NIH. Moss is a superb molecular virologist expert in several areas. In recent years he and also Enzo Paoletti of Virogenetics, Inc., in upstate New York, independently pioneered the use of vaccinia virus (smallpox) in an attenuated form to carry genes of other viruses. They succeeded in inserting the envelope gene of HIV isolates from my laboratory into vaccinia virus and in produc-

**Live attenuated virus* means a genetically modified virus that will not cause disease. This sometimes occurs after many passages through animals in a laboratory setting. Other times the "modified" virus was obtained in the field. For example, Edward Jenner's development of the smallpox vaccine apparently came out of observations that milkmaids in England did not develop the disease. He correctly reasoned that they might be protected because they were infected from cows with the cow pox virus, which cross-protected against smallpox but did not itself cause disease. *Killed whole virus* means taking a virus that causes disease and inactivating it (usually chemically). The "killed" virus may induce a proper immune response and not cause disease. Therefore, it could be protective. This is the case of the Salk polio vaccine.

†As in a virus that infects a bacterium and the inserted gene has signals that allow it to be active.

ing the HIV envelope protein from it. One can purify the HIV envelope from vaccinia virus–infected cell cultures and use it in the free form, as described in the first instance with a bacterium; or one can inoculate this vaccinia into people, as in conventional smallpox vaccinations, and in theory make a double vaccine.

Other investigators have used the same approach with polio virus vaccines using attenuated (non-disease-causing) polio virus strains. One theoretical advantage of this approach is obvious: the possibility of a two-for-one achievement. Another is not obvious. In a vaccination with live but attenuated polio or vaccinia, these viruses find their target cells and actively replicate in them for a period. In so doing their proteins will be released at the cell surface in a way and in a form to which the immune system has been designed through evolution to respond best. After all, this is the natural course of a viral infection. The HIV envelope protein delivered in this manner is likewise presented at the cell surface. In contrast, when we use a purified viral protein or a fragment of such a protein alone, it may not be presented in an optimum way. Consequently, the immune response may not be as great with this approach unless augmented by adjuvants.*

Some other difficulties in reaching a vaccine for HIV include the variation in HIV strains, the lack of a small animal model, and the probably unique need to prevent infection of every single virus particle because of the virus's capacity to integrate its genetic information into the host chromosomal DNA, and because of the probable quick and permanent infection of the brain where immune attack can be avoided. In the rest of this section I will summarize what has been accomplished and describe what I think are promising areas for the near future.

In 1985 my co-workers and I and several other groups from different countries formed an informal collaborative network, called HIVAC, for the purpose of exchanging ideas, information, and reagents. Our decision was to focus first on a subunit vaccine composed of a portion or all of the envelope protein of HIV. The desired immune response to viral proteins includes both antibodies, which can neutralize virus (block infectivity), and a T-cell response, which in-

*An *adjuvant* is the liquid material the vaccine is presented in and itself provides a stimulus to the immune system.

cludes T-cells that help B-cells make antibodies (some of the T4 lymphocytes) and T-cells that attack virus-infected cells (some T8-cells and others) as viral proteins are made and reach the cell surface, where they can be recognized by these T-cells. Although portions of most viral proteins can induce a T-cell response, in general, neutralizing antibodies are made only against the envelope of the virus. This has been the case with retroviruses.

Ideally, we would like to try every possible combination: whole viral envelope with and without its sugar side groups; the same viral proteins, each with different core proteins or with the reverse transcriptase or the regulatory proteins like *tat* and *rev;* and others. We would like to do all of these with different adjuvants. I am told that the adjuvant can make a great difference in the success of a vaccine. I believe this, but sometimes different adjuvants become the pet of different "vaccinologists" and can wind up sounding more like art than science. But the reader can see that the various permutations to test become extreme; finding the best combination will take time.

Since chimps have been the only available animal model that can be reproducibly infected with HIV-1 and since chimps are neither easily obtained nor, if obtained, easily utilized because of the often justifiable concerns of animal rights people, the difficulties for this program would have been severe from the start even if this were the only problem with an HIV vaccine. Probably such problems contributed to statements by some immunologists that a vaccine against HIV infection was impossible. Necessity forced us to the drawing board. My group would make available to a number of groups several different molecularly cloned genes of the envelope of different HIV isolates to make envelope protein. Different groups would take various approaches either to the method of production or the type of envelope made (whole or fragment, with or without sugar side groups, and so on). I have already mentioned one such early effort of Bernard Moss.

If we are limited to the chimp for infectivity studies needed for final definitive evidence of prevention of HIV infectivity, we are not limited in our capacity to study an immune response. Most animals should and do respond to the exposure to virus, purified viral proteins, or pieces of viral proteins by making antibodies and by cellular immunity even though their cells do not get infected by HIV. We could learn which approaches give the maximum response. Many scientists have contributed to this area, but critical to such early work were the groups at Duke University headed by Dani Bolognesi; a group from a small biotechnology company called Repligen, affiliated with MIT scientists and headed by Scott Putney; Marjorie Robert-Guroff, Marvin Reitz, and Peter Nara of LTCB; and the group at the Genentech Company. From these studies we learned which region of the envelope induces the maximum desired humoral immune response, that is, neutralizing antibody response. We also learned that the region could be isolated from the rest of the envelope and be just as effective, and that it induced neutralizing antibodies that were quite specific to the strain of virus used to make the envelope.

If we were to use this piece of envelope (and there are arguments for using components rather than the whole protein because some other regions might induce the wrong kind of antibodies, not only unhelpful but sometimes harmful), we would have to learn how to modify it chemically or what to add to it in order to broaden the neutralizing antibody response to cover all the various kinds of HIV strains that might infect a person. Also critical at the early stage was the work of Jay Berzofsky and his co-workers at the National Cancer Institute. We learned from Berzofsky some of the specific envelope regions that induce the kind of *cellular immune response* we are looking for, and again, the pieces work as well as the whole envelope. The proper combination of these pieces in the best possible preparation (adjuvant), either chemically modified or with other viral protein components to broaden the immune response, is in my opinion the heart of the future approach to an AIDS vaccine. It is where our efforts will focus. By these kinds of studies we can be more logical and selective and far less "shotgunning" (or at least we think we can) in the selection of candidate vaccines for studies in chimps.

We have also collaborated in cruder approaches designed to see what the whole envelope will do. One of the collaborating HIVAC groups, a vaccine company in Austria called Immuno is, in fact, now going forward with Phase 1 vaccine studies in humans. They made large amounts of the whole envelope from one of our strains of HIV. The whole envelope is called gp160.* After tests of the immune response to gp160 in small animals and some monkeys, they tested gp160 with their adjuvant preparation in a few chimps. They were

*Earlier I discussed that the HIV envelope, like that of any retrovirus, has two components, the outer larger-sized one, which for HIV has a molecular size of about 120,000 and has sugar-size chains, is called gp120. The inner portion is called GP41; the two together gp160.

sufficiently impressed by the response that they plan to test some of this material in humans in a pilot study. Such early studies are really designed to determine toxicity and to measure immune response by laboratory tests. Immuno has recently obtained approval for this, but the approach does not significantly differ from a study already performed in humans by a small U.S. company, MicroGene Sys. Its studies showed that the whole viral envelope was not toxic in humans. This agreed with less extensive but even earlier studies by Daniel Zagury but showed only a marginal and not very interesting immune response in these healthy volunteers, whereas Zagury's results showed a more impressive response.

While all of these studies were being carried out, a system was found that most of us believe will significantly advance the vaccine studies. It has been known since the work of Essex, Desrosiers, Kanki, and co-workers in 1985 that many African monkeys are infected with the simian immunodeficiency retrovirus SIV, which is about 40 percent homologous to HIV-1. Later it was shown by several investigators working in primate research centers that when some strains of SIV were inoculated into Asian monkeys (macaques) these animals developed AIDS, but African monkeys inoculated with the same strains did not develop AIDS—nor was AIDS ever documented in the various African monkey species that are the natural hosts of the particular SIV strain. SIV was isolated once from a macaque in captivity, but since this animal had been in contact with some African monkeys, and macaques are not known to be infected in the wild, it is assumed that the animal was infected in captivity from an African monkey.

So Asian monkeys are susceptible to SIV, as humans are to HIV. When HIV-2 was analyzed, it was found to be more closely related to the SIV found in this macaque. Now work by several groups (for example, Daniel Zagury in Paris, Gunnel Biberfeld in Stockholm, Patricia Fultz in Atlanta, and Genoveffa Franchini at LTCB) has shown that like SIV (but unlike HIV-1), some HIV-2 strains can infect and produce AIDS in macaques. With the assumption that experiments with SIV and HIV-2 in macaques are relevant to HIV-1 and humans, we now have a useful animal model without the severe constraints placed on us with the HIV-1 chimp model with its inherent disadvantages of lack of disease in the HIV-1–infected chimps and difficulty getting and working with chimps. Indeed, the most rapid and interesting advances in vaccine research are coming from primate centers using the SIV-macaque systems—centers such as those of Murray Gardner at Davis, California; M. Murphy-Corb at the Primate Center in Louisiana; Patricia Fultz in Alabama; and Ron Desrosiers at the New England Primate Center.

I think it was early in 1985 that, while in San Diego, I made an appointment with Jonas Salk. I saw him in his magnificent office with its vista of the Pacific Ocean, in the Salk Institute in nearby La Jolla. The Salk Institute is noted in biomedical sciences, mostly for its basic research. In some sense Jonas Salk seemed to be a fish out of water there, still far more interested in practical medical achievements than in the work that surrounded him. But that was exactly the kind of topic I wanted to talk to him about. I had already had much contact and many interactions with Hilary Koprowski, and would continue them, and to some extent also had input from Albert Sabin. Discussions with Jonas Salk would give me the benefit of learning from all three of those individuals, who, although not in harmony, nonetheless had done the most to develop the polio vaccine after the polio virus was isolated by Enders and co-workers.

Each of them has a different scientific outlook: Sabin is the most critical (sometimes it seemed for its own sake) and, in terms of identifying and isolating new viruses earlier in his career, one of the most successful virologists ever. Oddly, he seems the most pessimistic. Koprowski is the most responsive to doing laboratory experiments and is full of ideas and energy (he is the youngest). Salk is by far the most optimistic and dogged. Almost his first words to me were that we had the vaccine in our hands since that day in the fall of 1983 when we had succeeded in the continuous culture of HIV in cell lines. Salk believed then and continues to believe now in a whole killed virus vaccine approach. He formed a small vaccine company for the purpose of testing it. He has used mostly viral cores—killed virus, but without the outer portion of the envelope (the envelope is easy to lose in mass production of virus)—for experiments in monkeys, but also as a hoped-for immune stimulant against virus in people already infected with HIV, a form of immune therapy—as we have already seen.

He described some interesting early results in reducing and containing virus in a few infected chimps, and interesting and hopeful results in doing the same in HIV-1–infected people. As I was leaving his office, Salk remarked that up to then I was the only one who had sought his advice. Later, he was to be of help to me in other kinds of advice—as Hilary Koprowski has been for many years.

In contrast to use in immune therapy in a person already infected with HIV, I have emphasized my reluctance to consider a whole killed or an attenuated HIV vaccine in humans for use in prevention of HIV infection. Nonetheless, we all look with great interest for the results of these kinds of studies in the SIV macaque model at primate centers that have also used this approach, because even if not ultimately used in this form as a vaccine in humans, whole killed virus studies might tell us that a vaccine is possible with the SIV model in monkeys and presumably then also for HIV in humans.

As he was first to study immune anti-HIV therapy in infected people with vaccine-like materials, the first attempts to study the candidate HIV vaccines in humans for prevention were also by Daniel Zagury of the University of Paris, in close collaboration with clinical investigators from Kinshasha, Zaire. Zagury has been part of the HIVAC group from its beginning. We have collaborated with him in some of his work, as have other members of the HIVAC group, especially Bernard Moss, Dani Bolognesi and his colleague Tom Matthews, and Jay Berzofsky. Zagury's early aims were not unlike the later limited objectives of MicroGene Sys, Immuno, and Bristol-Myers: to determine the safety of purified viral proteins and to study the immune response to them first in animals and then in humans. Zagury's studies did differ from them not just by being the first such studies in humans but also, and more important, in producing the most interesting immune response in humans to date. In doing the studies first in himself, and in using a wider selection of vaccine components, Zagury has tested whole envelope with and without its sugar side chains, pieces of envelope with known capacity to induce neutralizing antibodies, and more recently some of these in combination with HIV-1 core proteins. He has also tested the presentation of HIV-1 proteins by different approaches. One of them employed the use of attenuated vaccinia virus containing the gene for the HIV-1 envelope protein.

His work is enhanced by closely knit collaborations with Zairian clinical investigators, but it is inhibited by the lack of sufficient HIV-1 proteins for larger studies. To summarize his findings: by attention to a program of periodic boosting of the immune response, he has been able to induce neutralizing antibodies and some cellular immune response. He has shown that the response can be maintained but in ways that may be impractical (frequent boosts have been needed). By injecting purified HIV-1 proteins or fragments of them into his skin, he has shown that some months after his first immunizations, he develops an impressive positive skin reaction. This is exactly the approach used to show that someone has been infected with the bacterium of tuberculosis. We call this test for tuberculosis the PPD skin test. It signifies not only a past exposure to the specific microbe but also that the person has immune memory to that microbe. None of these results, however, proves that Zagury or his other volunteers are protected against an HIV-1 infection, nor has he convincingly shown that the immune response he has generated is broad enough to cover the various HIV-1 strains.

Many other efforts—too many for all or most of them to be included here—are being made at developing an HIV-1 vaccine. I will mention just a few as particularly impressive examples. England has substantially funded an HIV-1 vaccine program, and recently Jeffrey Almond, Robin Weiss, and co-workers have reported early results of their approach using attenuated polio virus carrying the HIV-1 envelope gene. Alan and Sue Kingsman at Oxford University and their co-workers (also part of the early HIVAC collaborative efforts) have designed a novel approach. They found that yeast microbes can make particles similar to the cores of retroviruses. They inserted the HIV-1 envelope gene into the genetic region responsible for these particles. The advantage: presentation of a viral protein in a particle form is more in tune with the way the immune system is designed to respond, as I indicated in the discussion of its presentation in attenuated viruses or bacteria. Also, the yeast particles are innocuous but they are not infectious so they do not amplify the viral protein(s) as do some of the other carriers.

Wyeth Laboratories is using a harmless adenovirus to deliver HIV-1 proteins. The Merieux Company in Lyon, France, recently organized an extensive vaccine effort, as has the Merck Company. The Swedish program of SAREC, I have already mentioned as part of international efforts aimed at helping underdeveloped nations in problems created by AIDS, also has an HIV-vaccine development component. The Swedish program, headed by the immunologist Hans Wigzell, is an excellent example of a tightly coordinated effort focused on one major problem: the development of a successful vaccine in the apparently best animal model system, the HIV-2 infection of macaques. One leader, Gunnel Biberfeld, is responsible for the animal studies; another, Bror Morein, for adjuvants; Britta Wahren, for protein production; and others such as Erling Norrby, for defining the regions of HIV proteins most useful to focus on. Genentech Company has a program of testing HIV-1 envelope gp120 in chimps with some early impressive results, as does Mark Gerard and his co-workers in Paris.

The major positive results that developed, mostly in 1989 and 1990, were (1) successful protection in macaques vaccinated with killed whole SIV against subsequent challenge with infectious SIV, reported by most of the primate centers; (2) successful protection in macaques vaccinated with pieces of the SIV envelope against challenge with infectious SIV, reported from a collaborative effort of Walter Reed Hospital army investigators and NCI scientists and the Seattle Primate Center; (3) protection of two chimps vaccinated with HIV-1 proteins against subsequent challenge with infectious HIV-1 by Marc Gerard at the Pasteur Institute. The major problems remaining are how to maintain high-level immunity over a long period and how to solve the problem of virus variation—that is, the development of a vaccine for human use that will cover most if not all variants of HIV-1.

These multiple efforts in Europe and North America are extremely diverse. It should be clear that if a vaccine does succeed, in many ways it will belong to no one. Nor should it be named for anyone. The efforts of many have contributed the essential first steps. The efforts of many more will be needed for the essential final steps.

Epilogue

For the most part this book was written during the spring and summer of 1990. A few modifications were made during the fall months, but I wanted to add my last thoughts to it during the final days of the year. And so I have.

It is now the close of 1990 and I am heading for Thailand (my first trip to that country) to attend a meeting on AIDS called by Princess Chulabhorn Mahidol. AIDS is a relative latecomer to Thailand, as it is to most of Asia, but now Thailand joins the list of nations of major concern to the World Health Organization.

Just a few weeks earlier, I was at a similar conference in Ireland; here too the AIDS virus was relatively late in making its call but a sense of urgency now marks that nation's response. At the meeting in Ireland every possible prospect for limiting the spread of the disease and helping those already infected was examined, including the need to put in place the changes being made in other countries where the epidemic has struck: sex education of the public; vigilance in epidemiological surveys; testing blood for evidence of HIV infection prior to blood transfusions; appropriate early intervention in HIV-infected people with the available antiviral agents; and, possibly most important of all, offering understanding, information, and encouragement to HIV-infected people and to the public, both to protect infected people from discrimination and to help them retain faith in the biomedical research community.

This last will help those infected stay close to a qualified physician. What a terrible wrong would be done if infected people were to lose hope to the point that they rejected whatever medical help was currently available to them, especially as we approach the point where the medical community may actually come to have this disease in its grasp.

Even though HIV has by now spread to every corner of the globe, everything history has taught us about the never-ending battle between medicine and disease tells us that medicine may in the not-too-distant future begin getting the better of this new threat to world health and world prosperity. The best estimates at this time suggest that by the beginning of the new century, and maybe before, we should be able to offer substantial help to HIV-infected people through combination chemotherapy, something similar to the combination chemotherapy pioneered during the 1960s, which resulted in the cure of most cases of childhood acute lymphocytic leukemia.

AIDS will not necessarily be cured by such therapy. But with such treatment an infected person may be able to lead a normal life with some prospects for long-term survival, particularly if treatment is initiated early enough after infection. Such therapies may be augmented in time by a regimen in which pieces of HIV proteins are used to enhance the immune system's response to the virus. (Daniel Zagury, Odile Picard, and their collaborators in Paris, and more recently other groups as well, are testing just such a therapy.) Later on, gene therapy may also be introduced, and an effective vaccine to prevent against HIV infection may also be developed. It is still impossible to predict precisely when a successful vaccine will be available.

There is no shortage of good ideas for research on HIV. The question really comes down to which specific research attempts should be given priority, which is another way of asking who should be charged with the awesome responsibility of assigning such priorities.

HIV arrived and significantly spread in the United States sometime in the 1970s and was first clinically identified in 1981. I have often been asked what the prospects would have been for quickly identifying this disease's cause had it struck just a decade or two earlier (let us say in the 1960s). There is no way to answer that question with absolute surety, but the futility with which we might have been thrashing about is deeply unsettling.

My guess is that if extensive HIV infection had not occurred when it did, nobody in science would have thought to look for a retroviral cause; and if someone had, progress would have been painfully slow. Scientists simply did not believe that retroviruses caused disease in humans, and until the 1970s, no one had worked out the distinctive method by which animal retroviruses replicate. Nor had we developed the techniques necessary to study human retroviruses in the lab or to culture their target T-cells properly. In fact, we didn't even know then that T-cells and their subclasses existed. I suspect the cause would have been discovered, but the time required to do so would have been much greater and our understanding of what to do with that information far less.

As it was, we were fortunate that by 1980 we already knew enough about animal retroviruses—their genes, their proteins, their "life cycle"—and about human retroviruses as well—and were able to begin work on isolating the virus that caused AIDS and then on developing an effective blood test. Yet it remains sobering to note that even in 1980 there was still no concentration of talent in basic human virology, let alone human retrovirology, to which the worldwide human community could turn for rapid and reliable answers.

The AIDS epidemic also brought to our attention the fact that there may be a weakness in our biomedical system. Beyond the efforts of individuals or individual laboratories, which in the end did prove successful, there was no one unit with the primary responsibility for centrally coordinating AIDS research efforts. Not suprisingly, the first group to monitor the epidemic and to make available to those of us in research preliminary information about this new disease was the Centers for Disease Control (CDC) in Atlanta, with its notable experience in the study and control of infectious diseases. (Oddly enough, at just this time, there were political sentiments favoring the abolishment of the CDC, such was the level of false security we had then about our conquest of infectious disease.) But the CDC's expertise was in epidemiology and not so much in basic research, and there was no group of frontline research units designated to work in conjunction with the CDC to convert this preliminary information into advances in the laboratory.

This failure to marshal a centrally coordinated research assault to help find the causative agent, and to make sure that work began simultaneously on the opportunistic infections that accompanied AIDS, angered many, especially in the gay community, who could not understand how the wealthiest and most health-conscious nation in the world did not immediately undertake an organized effort. As a scientist, as a citizen, and from the vantage point of 1990, I must say that to a good measure I now believe their anger was justified, and this is something every thoughtful American should be concerned about.

Regarding the study of emerging human viral disease, scientists are united in the belief that such work is of scientific urgency. Regarding AIDS, we have two choices. We can set for ourselves the uncompromising goal of not stopping until we conquer this disease, or we can let it follow the path of previous microbes, in which case, even if we do nothing further, some equilibrium will eventually develop between us and the virus. When this happens, the *rate* of infection will level off on its own and then decline. This may result in a gradual acceptance of a new status quo, one in which there will be "tolerable" rates of infection for different regions of the world, or worse for some segments of a given population.

This has been the pattern in past epidemics, when science offered society no other choice. But even if we, as a society, were finally willing to accept this new situation regarding AIDS, it still does not offer us any comfort regarding the fact that AIDS may be the forerunner of other serious diseases caused by emerging viruses, each representing its own unique challenge. The scientific community is united in its belief that we must focus intensively on emerging viral diseases. I think, given enough information, the public would be also. The only unresolved question is how best to do this.

To ensure that a body of talent in virology is assembled and in place to maximize our efficiency in attacking AIDS and other viral diseases and to prepare ourselves for the next viral epidemic, I believe that at least one major international center of excellence for studies in human virology must be created in the United States, and a few more elsewhere.*

For such centers to solve the problem of new viral diseases at the pace we all want, they should be well funded, visible, and international in composition, and they should attempt to combine the best of philanthropy, industry, the university, and government. It goes without saying that until we make major progress with AIDS, this disease should remain the first and chief priority of such a center.

*In 1991, in Heidelberg, Germany, a new human virology institute headed by Harald Zur Hausen will officially open. However, it will focus solely on those viruses that appear to play a causative role in human cancers.

One might ask if such centers do not already exist as part of NIH, or should not be established under the existing NIH umbrella. Although work related to various aspects of human virology is, of course, now conducted at a number of NIH institutes, in point of fact there is no one institute devoted solely to human virology. I believe that a greater concentration of virologists at NIH would itself be helpful.

The question then arises whether such an institute should be established within NIH. There would be many advantages to doing so: its excellent international reputation, the opportunity it offers for contacts among a large pool of basic and clinical scientists in widely varying areas, proximity to Washington (hence to funds and political support as well as to numerous mechanisms already set up to conduct science). Nonetheless, this kind of decision must also be considered in light of the current and possibly future conditions at NIH overall.

As much as it pains me to say this, NIH is becoming increasingly disadvantaged in the competition for research talent. It has been hurt by the disparity in salaries and other income enhancements available to those in science outside government over the past few years, though I understand that a plan is now in preparation to help reduce the salary differences for the highest level scientists and clinicians. However, in my view there still remains the more important problem of the dangerously declining morale among scientists currently working there.

There will always be people ready to give up higher earnings for other satisfactions. Those who chose to work at NIH came here with a clear understanding of the trade-offs. They might earn less than in industry, but they were not restricted only to that scientific work that promised tangible practical results at a very early stage. Their flexibility was greater.

For many scientists the real question was whether to come to NIH or take a university position. At a university they would have the help of eager and talented students, greater salary flexibility, and the many other pluses of life at a university. At NIH they could expect long-term research support—they would not have to devote a substantial portion of their time to grant writing—and they would be virtually free of teaching responsibilities. Those who chose NIH and were accepted thought themselves very lucky indeed.

But in recent years, NIH, like many other institutions, has become more bureaucratized, sometimes offering enthusiastic scientists less intellectual freedom than they might obtain elsewhere, and increasingly asking many of its senior people, who came to do science, to place administrative chores and paperwork ahead of scientific productivity. Even more important, NIH scientists, as government employees, are subject to rules and checks that no other scientists must follow, and many of these rules come not out of science but of politics, a point I will return to in a moment.

There are other disadvantages as well to the idea of a new biomedical institute's falling solely under U.S. government auspices or, for that matter, directly under any one government's auspices. In domestic matters, government, all government, must usually move slowly, measuring every new step contemplated against the possibility of misinterpretation by powerful constituencies. As a consequence, embarking on any imaginative course of action requires of policy makers an ever more complex decision-making process. Unless our society, in general, finds some way to solve the problems of oversight and accountability without paralyzing the decision-making ability of managers, national institutes will, in my opinion, continue to decline in vitality, and will eventually become ill suited to addressing major scientific challenges, reducing them to dealing only with mundane tasks.

But the strongest argument for creating the kind of centers I propose is a positive one: they will take the best of government, the university, and industry and combine them, while resisting the encroachments of any one of them. Government involvement is essential for funding and political support; university involvement adds needed relationships with young scholars and an in-depth interaction with talented faculty, and appeals to those many scientists who feel a need for a university appointment; industry offers the needed practical, technological help—bringing basic observations to the next levels. Some have told me this is utopia. It would not be politically possible. But I do not know why. In fact, under the past NIH director, James Wyngaarden, we almost succeeded in doing precisely what I am recommending right on the NIH campus.*

*A private foundation was prepared to provide funding for the buildings and equipment needed and even for start-up costs. There was interest by some major U.S. universities, institutes in Israel and Europe, and, of course, by industry. By the time the agreements were made there were financial changes which led to a change in the decision or capabilities of the foundation. If it had gone forward then mechanisms for being sure there would not be conflicts of interest would have to have been carefully installed, but I do not see why it cannot be done. Some other nations seem to have worked out such combinations.

For all these reasons I think a free-standing international human retrovirology center is the next logical step. But I would be less than fully honest if I did not admit—to myself as well as to readers—the extent to which my thinking has been influenced by a series of events of the past few years, events I would not have dreamed possible during many years as a cancer researcher. These events have dramatically changed my life as an NIH scientist.

I have debated with myself whether to use these pages to discuss any aspect of these matters, knowing that to do so will detract from the science, move the reader away from the broader problems, and focus on my own. I have concluded that some of the questions they raise are larger than my personal experience and that it is important to get other scientists, as well as the public, thinking about those aspects of these problems that threaten to change the way scientific work is conducted, with consequences no one can predict. And so I have decided to talk about the newspaper article that led to an inquiry by the NIH into my lab's AIDS research.

Most medical scientists of reasonable productivity find at some point in their lives that the work they are doing is of immediate interest to the public. Interest in science has increased enormously over the past two decades; articles published in the scientific journals are regularly picked up and carried as front-page news items in the nation's daily newspapers, often within days of their scientific publication. During most of the twenty-five years I have been at NIH and until very recently, I have tried to accommodate reporters' requests for interviews if I thought I could be helpful. I have always assumed that there was a sincere interest on the part of the reporter in writing a good and fair story. Most of the time I have not been disappointed.

It was in just such a spirit that in the late 1980s I spoke with a reporter from the *Chicago Tribune* by telephone about our AIDS work. It was soon clear to me—by the questions asked—that his attitude was not that of the independent investigative reporter trying to understand science. Instead, I had become the target of a reporter with a mission, a reporter who somehow turned from exposition to exposé, from analysis to assassination. In so doing this reporter rejected what I naively thought at the time to be the most important traits any journalist brings to a story: honesty and a sense of fair play.

Had I been a scientist at a university or in private industry, the story might then have taken a difference course, because in either of these two cases, neither I nor my colleagues would have had the obligation to turn written records of our work over to a member of the press. But as government employees, my colleagues and I are subject to the provisions of the same Freedom of Information Act, which makes available to those who request them—journalists, scholars, the public—the records (except some few classes specifically excluded by the provisions of the act) of all government agencies, the NIH included. (For example, the lawyers representing the French government in their blood test patent suit also used FOIA to obtain information from our lab.) This reporter therefore had the right to request and receive virtually all my records as an NIH scientist, including requests for much of my travel records, my correspondence with any other scientist, internal memoranda I wrote, any records my colleagues and I kept on the workings of the lab, and other documents. Using this statute, he did just that, making literally hundreds of FOIA (Freedom of Information Act) requests for documents regarding me and/or my colleagues, obliging us regularly to photocopy and send out to him almost anything that suited his whims.

Of course, it should be obvious that non-U.S. government scientists are under no such similar obligation (no other country in the world, as far as I know, has a Freedom of Information Act covering government-based scientists), and so none of their records were ever obtained, except those they themselves decided to make available. Even more important, although the Freedom of Information Act required me to release most of the documents requested, nothing in that act required the reporter to incorporate into his story information obtained from those documents that contradicted his own interpretation of events. In other words, I had to make everything available but the writer could pick and choose what he wanted.

It doesn't take much imagination to see that the freedom to *select* information becomes a license to disseminate *mis*information. Scientists I have spoken with are surprised that those outside the scientific community don't understand this as clearly as those within science do. Scientists know that an investigator who incorporates into his or her conclusions only those findings that support the hypothesis produce work of absolutely no probative value. Put most broadly, given the freedom to choose which findings to include, a scientist could prove that the sun goes around the earth.

As the pattern of this reporter's investigation became apparent, we tried to reach both the writer and his editors, but when the article came out it was a mockery of fair journalism. On a most basic level, there were instances where the information was simply not correct. (I am told that this is the norm in most general-interest journalism and often occurs without ill will on the part of the reporter. In most stories it is not easy to get everything right, and reporters are often rushed to meet deadlines.) But more disquieting, the information in this article was slanted as presented and its significance examined only in the one light—the one that favored the journalist's thesis. Statements from scientists and other sources were screened to include only those fragments that reinforced the author's point of view. The great writer Miguel Cervantes said that the half-truth was the worst form of slander. I now understood what he meant.

The article came out at a key point in our research on AIDS. Both our lab and the Pasteur Institute scientists believed we had put our disagreements behind us with our signed agreement, an agreement that involved not simply money for an international AIDS fund but a written record of the accomplishment of both teams. The agreement had one other major advantage as well. It had elicited the wholehearted support of the scientific community, whose interest was in seeing science get past this politicized controversy and back to work on the problem. Now here was this reporter recharging the atmosphere.

There is much irony in all of this. It is the U.S. government that gave the world the first notion of continuous long-term major support for basic biomedical research. This support laid most of the groundwork for the advances of mid- to late-twentieth-century medical research; this support made possible the groundwork science that led to the work of my lab. For over two decades this support made the work of my lab possible, which, in turn, led to the discoveries of human retroviruses, to some of the technology to study them, to the evidence that convinced people that HIV was the cause of AIDS, and to most of the work leading to the development of a workable and convincing blood test for the AIDS virus, which saved so many lives and halted a far greater spread of the virus than would have occurred without it.

Thus, I did not question the government when it made its request of me and my co-workers to patent the blood test. (I was told it was important to encourage the larger pharmaceutical companies to enter into licensing agreements; only they had the resources to move quickly and effectively.) However, when the Pasteur Institute instituted its suit, its strategy was to wage a public relations campaign that focused first on their contention that it had been denied full recognition for the first isolate of an AIDS virus and second on me personally as a visible target.

Unfortunately, no one in the U.S. government came forward to make clear that I had not requested this patent, and that, as the law stood at that time, I had no reason to believe I would ever financially benefit from it. As far as I knew, no NIH scientist had ever benefited from a patent based on his or her work.

Now I was also left unprotected by the terms of the Freedom of Information Act. Though I am as proud as the next person that the United States believes in freedom of information, and though we have never failed to cooperate with any legitimate inquiry into our lab results, the FOI statute is capable of being turned into an instrument of personal harassment. When it is, there is little the victim of such harassment can do to stop it, particularly if he is a public figure.

In November 1989 the *Chicago Tribune* gave over fifteen standard-size pages to their story. When all was said and done, we were back to the same old story raised years earlier by litigation lawyers and their public relations firms prior to our agreement with the Pasteur Institute—whether I or someone else in my lab, either through direct misappropriation or accidental contamination, had inappropriately put to use the final LAV samples given us by the Pasteur Institute during our AIDS research.

Of course, this is not the first time someone publicly visible in his or her field has fallen victim to unfair reporting. The big problem with the *Chicago Tribune* story is that it interwove a blatant personal attack on me with an appeal to antiscientific attitudes, to the notion that scientists cannot be trusted, that we feel we are above the law, and that I would sacrifice my integrity and honor for notoriety. Yet, despite the promises of groundbreaking revelations, this massive article failed to come up with any charge or accusation capable of holding up under scrutiny. The charges rippled through the scientific and public communities and died. Some few months later (sometime in early 1989), however, an unprecedented series of events occurred to resurrect the piece.

First, many hundreds, perhaps thousands, of copies of the article were sent, anonymously, to my scientific colleagues and collaborators and to scientists and science administrators all over the world. Many mailing lists must have been used, for one European scientist told me he received no fewer than four copies mailed to his home address.

I have asked other reporters about this practice and they tell me that it is not uncommon for a reporter who has written a "big" story to send copies of it out to others in the media to encourage them to reprint or quote from the original story. But in this case, the piece was not being sent to other members of the press, where it might have done this reporter some good, but to other scientists, where it was apparently hoped that it would do me some harm.

But while the mailing surely broadened the sheer number of scientists aware of the article, it did not give further legitimacy to it. To their credit, the U.S. scientific press, most of the world press, and the major scientific journals did not in any significant way acknowledge it. It was the next part of the campaign that had the first real major impact on my work.

NIH, as part of the Department of Health and Human Services, is overseen by a committee chaired by Congressman John Dingell (D-Mich.). Over the past few years, there has been a number of alleged instances of scientific fraud and/or misconduct that have caught the attention of this committee. Some have been shown to be baseless; others borderline; a few clearly of concern. In the last category were cases in which the scientist had simply not performed the experiments he or she claimed to have performed or the sample used had been grossly exaggerated, and thus the findings were fraudulent.

In most (perhaps all) of these cases, other scientists first brought these fraudulent experiments to the attention of administrators at the facilities where such experiments were said to have been performed. This committee became involved because most of these experiments, though done elsewhere, had been NIH-funded.

In recent years, a number of these inquiries into claimed, presumed, or substantiated scientific misconduct gained media and congressional attention. At congressional hearings at least one scientist reported from personal experience that NIH overseers had been less than vigilant in investigating these incidents of fraud (in at least one instance investigating the whistleblower rather than the one accused of fraud). As a consequence, during such hearings the question was raised as to whether the system that had long obtained—that which allowed scientists to oversee the conduct of other scientists—should be replaced by one in which lay people played more of a role. The rumors around NIH were that Dingell's committee was eager to have such questions resolved by nonscientists under congressional auspices.

Controversy is no newcomer to science, nor are occasional incidents of scientific fraud. Brother Gregor Mendel, the father of genetics and a cleric, has been accused of having fudged his pea data; his critics say that it is highly improbable that he could have come up with such precise findings with the limited number of peas in his study. Was Mendel dishonest or just very lucky? We do know he was right and his contributions were enormous.

I have even read reports of controversy about the work of the great physicist Isaac Newton, although here, too, no one has challenged, or can ever challenge, his contributions to science. Quite recently it was reported that the great German scientist Johannes Kepler, who described the elliptical orbit of planets, completed a table (accurately) with data derived not from his raw initial observations, as one looking at the table would be led to believe, but from later experiments. Would this be viewed today as a scientific peccadillo or an example of punishable scientific misconduct?

Of course, there are also scientific experiments that were challenged that did prove in fact to be fraudulent, doing damage to a scientific community that presumes data to have been arrived at honestly. One that is often mentioned with confidence is that of the English psychologist Sir Cyril Burt, who reported truly remarkable IQ similarities among fifty-three pairs of identical twins raised apart, giving enormous support to the nature side in the nature-nurture dispute. Then it was revealed that Burt had perpetuated a massive fraud on his colleagues and the public, and that his investigation did not follow the methodology he claimed it had. (I was certain when I chose this last example that here was an open-and-shut case of scientific fraud, only to open this week's issue of the journal *Science* [January 4,1991] to read that at least one person believes Burt's work may yet be vindicated.)

Scientists justified the old system on the unspoken premise that they had the most to lose by not effectively policing their own. After all, science is a cumulative process. The surgeon may bury the evidence of his own malpractice, but the scientist must turn his results over to colleagues to examine, attempt to reproduce, go forward from. One scientist builds on the work of another and if scientist B cannot rely on the findings of scientist A, he or she cannot build on those findings with any assurance that his or her own work will rest on a solid foundation. Therefore the fate of most incorrect work is simply to die out.

But there was something else just as disturbing about the idea of moving control of cases of scientific fraud out of the domain of other scientists. This change seemed to be part of a larger change slowly revolutionizing laboratory science.

The proposed new "lay" oversight was not really lay—no one was talking about a diverse cross section of the general population—but a clear shift in the control of science from the body of working scientists to a new group of "experts."* Some ethicists are trained totally outside science; others are scientists who found a new niche for themselves in the "ethical" oversight of their peers instead of in laboratory or clinical research, in clinical medicine, or in teaching and administration.

And at a more intellectual level there was an even more important shift. It used to be that the measure of a successful experiment was the fact that your colleagues could replicate your findings independently. Now a piece of research under attack would have to meet a new standard of proof, one that did not really evolve out of science—namely, that the researchers had recorded and preserved the data in a form satisfactory to an independent reviewer who may (and in fact usually does) conduct his investigation several years after the experiments took place. Such a change, of course, rewards the more meticulous scientist and penalizes many less careful but nonetheless quite creative and productive scientists who, from my own experiences, are every bit as important as the former in advancing scientific knowledge.

It would be in the context of all of these changes in science that the *Tribune* piece, though it had little appeal within the scientific community or the scientific press, next showed up in the hands of Congressman Dingell's committee, prompting a letter from him to NIH, asking the acting director William Raub† if NIH was aware of the article and its innuendos. No one ever challenged the reproducibility of our scientific findings. The situation was more complex.

*The application of bioethical principles to the conduct of clinical research or the practice of medicine, such as the requirement that researchers and attending physicians obtain the informed consent of subjects/patients, is a separate matter not treated in this book.

†Dr. Raub has been acting director ever since the early period of the Bush administration. No permanent director has been appointed as of this writing.

Implicit in this request for a response from the NIH was the question of whether NIH really did do an effective job of overseeing issues of scientific fraud, and whether NIH would be as fastidious in pursuing accusations of wrongdoing against one of its own labs as it claimed it was in pursuing accusations of scientific fraud in those whose work it funded. NIH administrators presumably felt there was no choice but to demonstrate that NIH would and could pursue one of its own, and would do it even more aggressively than it pursued scientists outside NIH.

I was shocked. I argued that none of the pertinent points in the article were novel; charges of wrongdoing had not actually been made, only hinted at; they had all been dealt with in our 1987 agreement with the Pasteur Institute, a fact both the Pasteur administrators and their lawyers acknowledged. I asked if the article had been peer-reviewed or in any way verified by outsiders for accuracy. I stressed the fact that the piece of "evidence" used as one of the linchpins of the article was the striking similarity of one of our HIV isolates, the 3B strain, to the LAV/BRU strain from the Pasteur Institute scientists. But we had long ago noted and published these findings. Further, I clearly had the records to show that it was we who first noted the similarity and reported it to the French. Most important, it was clear even then that we had no need to misappropriate the Pasteur virus sample, since we were free to work with it and had isolates of our own to work with, including the first ones cultured in cell lines.

Over time, our position on all matters of original contention was borne out. But once started, inquiries develop a survival instinct all their own, and in a second phase, I was asked to review for the committee certain points in one of our articles, regarding matters that I believe to be quite technical in nature. In a single meeting all remaining questions were answered candidly and, in my mind, definitively.

Will this be the end of things? I continued to hope so.

One final point: last year I was asked to give a lecture at Bard College in upstate New York. My lecture was preceded by an introduction by the host that was typical in most respects, but was also warmer, friendlier, and more considerate than I would have expected from the person making it, solely because we had not previously met. Late in the day I returned to New York City by car with my host and used the opportunity to express my thanks for his gracious reception of me.

He smiled and slowly replied that he felt we had some "kinship." I wondered why and asked. He seemed at first reluctant to talk about whatever was going through his mind but eventually he told me that an individual who said he was associated with the Dingell committee had called him more than once to suggest that he "disinvite" me. That individual, I learned, stressed his "close knowledge" about my NIH inquiry.

This was not the first time an attempt to hurt me had been made, but what the caller playing the Roy Cohn role did not know was that this time he was speaking to a person who could not be reached by such tactics. Before it was fashionable to be a target of McCarthyism, my host lost his job at Princeton University's Center for Advanced Study after just such an experience. His position there had been as Albert Einstein's mathematician.

As I end this book, I am reading the final version of an article about to be submitted to a scientific journal. I can now say that because of the inquiry, we have undertaken a new evaluation of the question of the origin of 3B, our blood test isolate. I will not discuss the content of this article here because it has not yet been published in the peer-reviewed scientific literature. But for those readers who maintain an interest in the origin of this virus isolate, some answers will be provided when this paper and others are published in the spring/summer of 1991 and perhaps there will be some additional questions as well.

But the origin of 3B or LAV/BRU was never a burning issue to those with a true interest in science. I know the Pasteur Institute scientists did indeed have a bona fide HIV isolate in 1983. I know we did too. I know and I have demonstrated that we also had several others readily distinguishable from 3B (by molecular analysis), therefore, independent of 3B. Thus, the issue has no real scientific interest, little historical importance, and should not affect what is usually called "credit." I cannot believe the question is of interest to those infected with AIDS still waiting for a cure or to those not yet infected but hoping to see a vaccine produced so that future generations will not be harmed by the virus. It is, at best, of interest only to those who are fascinated by historical virology.

The 1983 work of the Pasteur group driven by Luc Montagnier, Jean-Claude Chermann, and Françoise Barré-Sinoussi and helped by some clinicians like Willie Rosenbaum and other virologists, such as Françoise Brun-Vezinet, was and will always be credited with contributing seminal advances in the field. The agreement between us and our governments over the division of scientific credit endures—despite all the attempts of an obsessed reporter to provoke one of the sides to reject it.

It is also important that questions of scientific wrongdoing be dealt with. Of equal importance is that scientists not be asked to respond to politically inspired inquests.

What is most important now is that science move ahead, that we develop an AIDS vaccine, that we help fight the opportunistic infections that currently invade AIDS-infected patients, and that we develop more and better anti-HIV agents.

In the two and one-half decades I have been in biomedical research, I chose to stay closer to the laboratory rather than to go to higher posts, thus avoiding larger administrative roles. These years have often been exciting, filled with unexpected surprises and rewarding experiences. Admittedly, there have been many twists and turns, ups and downs, but to me this is the character of life in science. I hope it will remain so. With some help to avoid overbureaucratizing scientific research, I am sure that the next decade will also be one in which we will have opportunities to make many more observations and discoveries, and that these basic laboratory findings in AIDS research and in cancer research will be translated into advances in treatment and prevention in the clinic. At the same time, we must be sure not to cut off the wellspring of our achievements, the support of basic research.

As we look back, it was the decade of the 1970s when we acquired both a fundamental understanding of animal retroviruses and the knowledge to develop the technology to isolate human retroviruses. At the beginning of that decade, one scientist, trying to understand the replication cycle of the Rous sarcoma virus of chickens, made the creative leap that resulted in the discovery of reverse transcriptase. This enzyme was soon shown to be carried by all retroviruses infecting animals. As important, it was shown to be the key to the novel mechanism by which retroviruses successfully infect cells and thereby replicate themselves and cause disease.

In time, the discovery of this one enzyme opened the door to new technologies in molecular biology and provided us with a new and important tool for virus hunting in humans. For once we knew about reverse transcriptase, instead of beginning our search for evidence of the elusive human retrovirus by finding free-standing viral particles, we could simply look for evidence of this distinctive enzyme as a footprint, or marker, of the presence of a retrovirus.

It was also during this period that our attempts to grow normal and malignant blood cells led to our finding the important T-cell growth factor Interleukin-2. Now we could routinely grow T-cells in the laboratory, a breakthrough that proved critical to our success in finding the first human retroviruses. But Interleukin-2 proved to be a discovery with enormous potential beyond the uses we had imagined for it, for it helped open a new era in T-cell immunology in general.

Also critical to our work in this period were the key advances made by others, particularly immunologists, in distinguishing different types of blood cells, especially subtypes of T-cells. For example, monoclonal antibodies were discovered during these years. Their discovery greatly facilitated our ability to subtype various human blood cells, including the important T4 lymphocyte.

By the end of the 1970s, with equal measures of persistence and luck, we were able to find the first human retroviruses, HTLV-1 and HTLV-2. Work done halfway around the world in Japan, by an innovative hematologist, led us and Japanese scientists to identify this new human virus in not just a few but many patients with a particular form of leukemia. Additional studies conducted soon thereafter in the Caribbean islands gave us further evidence of this same correlation. For the first time there was a clear association between a retrovirus and a malignant disease in humans, and a method for studying both in the laboratory.

At the start of the 1980s, we could not have anticipated that the cause of a new and frightening epidemic disease soon to be known as AIDS would turn out to be a third human retrovirus. But scientifically at least, enough preliminary work had been done to allow us to test for a possible retroviral agent. In two years we found the AIDS retrovirus. Once again, amazingly, it turned out to infect T4 lymphocytes. Thus, the 1980s turned out to be a decade of awareness and understanding of how retroviral disease attacks humankind.

So, finally, we are facing the decade of the 1990s with a broadened base of knowledge. It is my hope that this knowledge will contribute to clinical advances that will allow us to treat AIDS patients and certain cancer patients more effectively and to develop vaccines for the once-mysterious family of viruses called human retroviruses.

The AIDS virus has had much time to improve itself biologically in order to escape the immune system of those persons it infects. So far it has gotten the better of us. My hope, my belief, is that in the next decade or so we will find a way to turn the tables. This is the kind of battle and the kind of competition we like to be in. I am excited by the prospects for success.

Name Index

Subject Index

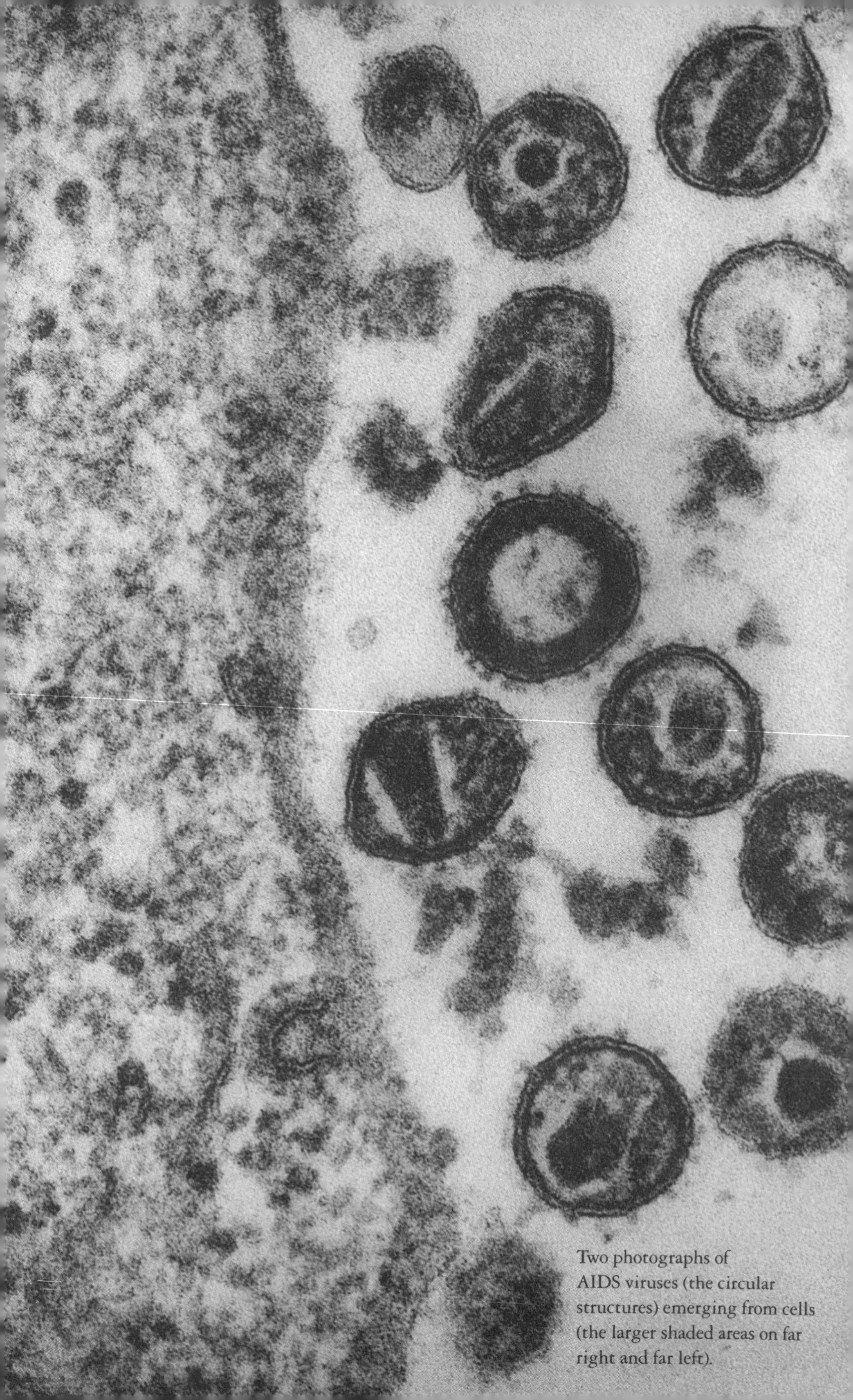

Two photographs of AIDS viruses (the circular structures) emerging from cells (the larger shaded areas on far right and far left).